Matrix Transforms for Computer Games and Animation

John Vince

# Matrix Transforms for Computer Games and Animation

Professor Emeritus John Vince, MTech, PhD,
DSc, CEng, FBCS
Bournemouth University
Bournemouth, UK
www.johnvince.co.uk

ISBN 978-1-4471-4320-8 ISBN 978-1-4471-4321-5 (eBook)
DOI 10.1007/978-1-4471-4321-5
Springer London Heidelberg New York Dordrecht

Library of Congress Control Number: 2012942604



Printed on acid-free paper

Springer is part of Springer Science+Business Media (www.springer.com)

*This book is dedicated to my best friend, Heidi.*

# Preface

This is an introductory book on linear matrix transforms, and should be of interest to animators and programmers working in computer games and animation. Although I have covered many of the topics in other books, I have never addressed the subject of matrices as an individual topic—hence the reason for this book.

The book's structure is very simple: Chap. 1 provides a short introduction to the book's objectives. Chapter 2 gives the reader some historical background and algebraic evidence for the matrix as a valid mathematical object, and its associated scalar-valued determinant. Chapter 3 describes how the determinant is computed for different size matrices. Chapter 4 provides a formal description of matrix algebra, with plenty of worked examples. Chapters 5 and 6 describe 2D and 3D transforms respectively, again with plenty of worked examples. Chapter 7 provides an introduction to quaternions with an emphasis on their matrix formulation. Finally, Chap. 8 concludes the book.

I would like to thank Beverley Ford, Editorial Director for Computer Science, and Helen Desmond, Associate Editor for Computer Science, Springer UK, for their continuing professional support.

Ringwood, UK John Vince

# Contents

# Chapter 1
# Introduction

## 1.1 Matrix Transforms

Ever since the invention of computers there has been an acute interest in using them for graphical applications. In the first computers, when cathode ray tubes were used to display messages, it was possible to manipulate patterns of characters and simulate simple games such as 'noughts and crosses'. When vector graphic devices were invented, lines could be drawn on display screens and used to represent 2D plans and eventually perspective views of 3D objects. Computer graphics was eventually born from these humble beginnings, and today it has become a global industry, and so large that is difficult to estimate its true value. Computer games and animation are two important areas of computer graphics, and during the past fifty years software systems have been developed to model, animate, render 3D scenes and create life-like characters.

From small graphic icons on mobile phones, to the complex CAD descriptions for the Large Hadron Collider in Geneva, computer graphics has emerged as a sophisticated science that relies upon mathematics to describe its algorithms. Mathematics plays a central role in representing geometry, colour, texture, illumination and motion. Simultaneously, hardware in the form of high-performance graphics boards, have become so fast and sophisticated that gaming systems costing a few tens of dollars, can render complex scenes in real time with incredible realism.

No matter whether computer graphics is used in medical imaging, graphing stock movements, film special effects or computer games, one finds a common set of mathematical tools in the form of matrices. These mathematical objects provide a convenient way to control scale, position, shear, reflection and rotation. However, they are also widely used to generate curved lines and surfaces. But in order to describe the role of matrices for curved lines and surfaces, one would have to cover the algebra of quadratic and cubic curves, which is described in detail elsewhere. (*Mathematics for Computer Graphics* by the author.) So, in this book, I have confined the application of matrices to the above mentioned transforms.

Out of all the transforms used in computer graphics, rotation causes most problems. The reason for this is two-fold: First, rotations can be difficult to visualise; sec-

J. Vince, *Matrix Transforms for Computer Games and Animation*,
DOI 10.1007/978-1-4471-4321-5_1, © Springer-Verlag London 2012

ond, they are frequently combined to rotate about two or three axes. Consequently, I have described rotation transforms in some detail, and I have also spent some time describing rotation about an arbitrary axis. I have also included material on quaternions from my recent book *Quaternions for Computer Graphics*, as they have a useful matrix representation.

## 1.2 Mathematics

In mathematics one often comes across statements such as: "By definition, the determinant of a matrix is given by...." or "By definition, a quaternion is an ordered pair...." Personally, I find such statements annoying, as they fail to explain why such definitions are made. It is unlikely that a mathematician suddenly thinks "I am going to invent matrices which have a scalar value called a determinant, calculated as follows." Or, "I am going to invent quaternions, whose magnitude is computed by...." No, the reality is that after many months of analysis and trying to understand a subject, a mathematician identifies the salient features of an invention and formalises various definitions in order to accurately document its characteristics. In the case of quaternions, Sir William Rowan Hamilton spent many years trying to generalise 2D complex numbers to a higher dimension, and eventually stumbled across a four-dimensional object comprising a scalar and a 3D complex number. A quaternion is not an arbitrary definition by Hamilton, but something that emerged through years of continuous study and research, until a spark of genius inspired him to invent the object we know today.

One can understand the need for definitions in mathematics, for without them there would be anarchy. Furthermore, it would be impractical to explain the history behind every mathematical invention when introducing a new topic. However, as this is an introductory book, I have tried to explain why various definitions are made, which hopefully will improve the reader's overall understanding of matrices and their role in transforms.

## 1.3 The Book's Structure

Matrix notation is introduced in Chap. 2, where I have shown how the notation emerges naturally from the algebra associated with solving simultaneous equations. One of the important expressions found in the algebra of simultaneous equations is represented by the determinant and is closely associated with a matrix. Chapter 3 shows different ways to compute the determinant. Having shown how matrix notation emerges from everyday algebra, Chap. 4 describes it formally, covering the zero, negative, diagonal, identity, transposed, symmetric, antisymmetric, orthogonal and inverse matrix. 2D and 3D transforms are described in Chaps. 5 and 6 respectively. Finally, Chap. 7 introduces quaternions as an object for rotating vectors about an axis, and develops their matrix form.

# Chapter 2
# Introduction to Matrix Notation

## 2.1 Introduction

In this chapter matrix notation is introduced as a tool for solving a pair of linear equations. This reveals three important features about matrices: The first is the existence of a scalar value associated with a matrix called the *determinant*; the second is *matrix multiplication*, and the third is *matrix inversion*. This prepares us for the next chapter where we investigate the nature of the determinant for larger matrices.

## 2.2 Solving a Pair of Linear Equations

There are two simple ways to solve a pair of linear equations such as

$$24 = 6x + 4y \tag{2.1}$$

$$10 = 2x + 2y. \tag{2.2}$$

The first technique is graphical and the second is algebraic.

### 2.2.1 Graphical Technique

The graphical technique represents the equations as two straight lines which may be coincident, parallel or intersect. The point of intersection could be located anywhere with respect to the Cartesian axes, and could make it extremely difficult to identify an accurate $x$–$y$ position.

Figure 2.1 shows two lines representing (2.1) and (2.2), where the solution is the point of intersection $(2, 3)$. Although the solution is easy to identify for these equations, a more reliable and accurate technique is required. So let's consider an algebraic approach.

J. Vince, *Matrix Transforms for Computer Games and Animation*,
DOI 10.1007/978-1-4471-4321-5_2, 

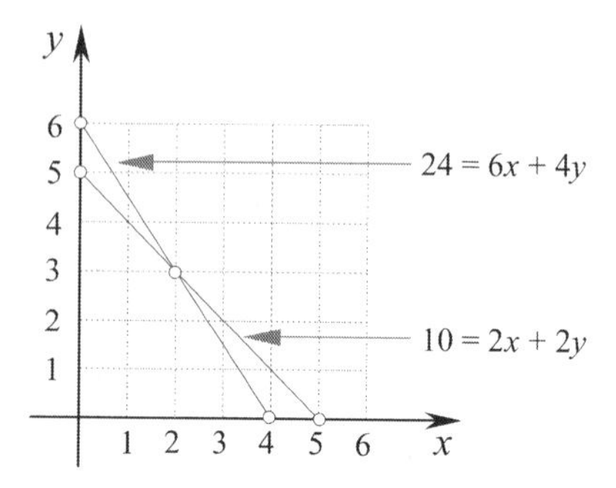


**Fig. 2.1** Graphs of the simultaneous linear equations

### 2.2.2 Algebraic Technique

The algebraic strategy is to manipulate (2.1) and (2.2) such that when they are added or subtracted, the $x$ or $y$ coefficient disappears, which permits one variable to be identified. The second variable is revealed by substituting the first in one of the original equations. We begin by multiplying (2.2) by 2 to turn the $2y$ term into $4y$:

$$2 \times 10 = 2 \times 2x + 2 \times 2y$$
$$20 = 4x + 4y.$$

The pair of equations now become

$$24 = 6x + 4y \tag{2.3}$$
$$20 = 4x + 4y. \tag{2.4}$$

Subtracting (2.4) from (2.3) produces

$$4 = 2x,$$

which means that $x = 2$. To discover the corresponding value of $y$, we substitute $x = 2$ in (2.1) to discover that $y = 3$.

This algebraic approach always provides an accurate result, so long as the original equations are linearly independent. Now let's find a general solution for any pair of linear equations in two unknowns, with the proviso that they are linearly independent. We start with the following pair of equations:

$$r = ax + by \tag{2.5}$$
$$s = cx + dy. \tag{2.6}$$

To eliminate the $y$ coefficient we multiply (2.5) by $d$ and (2.6) by $b$:

$$dr = adx + bdy \tag{2.7}$$
$$bs = bcx + bdy. \tag{2.8}$$

Next, we subtract (2.8) from (2.7):

$$dr - bs = (ad - bc)x,$$

and

$$x = \frac{dr - bs}{ad - bc}. \tag{2.9}$$

To eliminate the $x$ coefficient we multiply (2.5) by $c$ and (2.6) by $a$:

$$cr = acx + bcy \tag{2.10}$$

$$as = acx + ady. \tag{2.11}$$

Next, we subtract (2.10) from (2.11):

$$as - cr = (ad - bc)y,$$

and

$$y = \frac{as - cr}{ad - bc}. \tag{2.12}$$

Note that (2.9) and (2.12) share the same denominator, $ad - bc$, where $a$, $b$, $c$ and $d$ are the coefficients of $x$ and $y$ in the original simultaneous equations. This denominator becomes zero if the equations are linearly dependent, and no solution is possible.

Let's test this general solution using the original equations (2.1) and (2.2) where:

$$a = 6, \quad b = 4, \quad c = 2, \quad d = 2, \quad r = 24, \quad s = 10$$

$$x = \frac{dr - bs}{ad - bc} = \frac{48 - 40}{12 - 8} = 2$$

$$y = \frac{as - cr}{ad - bc} = \frac{60 - 48}{12 - 8} = 3.$$

The algebraic solution reveals some interesting patterns which emerge when we consider a third technique using matrix notation, which is covered next.

### 2.2.3 Matrix Technique

Given a pair of linearly independent equations such as (2.1) and (2.2):

$$24 = 6x + 4y$$

$$10 = 2x + 2y$$

their solution must depend on the constants and coefficients 24, 6, 4, 10, 2 and 2. Matrix notation describes equations such that the coefficients are isolated from the variables of $x$ and $y$ in a $2 \times 2$ *square matrix*:

$$\begin{bmatrix} 6 & 4 \\ 2 & 2 \end{bmatrix},$$

or in the general case:

$$\mathbf{A} = \begin{bmatrix} a & b \\ c & d \end{bmatrix},$$

where $\mathbf{A}$ identifies the matrix. The denominator $ad - bc$ in (2.9) and (2.12) is the difference between the cross-multiplied terms $ad$ and $bc$. This is called the *determinant* of the matrix, and is written.

$$|\mathbf{A}| = \begin{vmatrix} a & b \\ c & d \end{vmatrix} = ad - bc.$$

To keep the notation consistent, the values 24 and 10, or the general values $r$ and $s$, are represented as a *column matrix* or *column vector*:

$$\begin{bmatrix} 24 \\ 10 \end{bmatrix} \quad \text{or} \quad \begin{bmatrix} r \\ s \end{bmatrix}.$$

The variables $x$ and $y$ are also represented as a column vector:

$$\begin{bmatrix} x \\ y \end{bmatrix}.$$

Finally, we bring the above elements together as follows:

$$\begin{bmatrix} 24 \\ 10 \end{bmatrix} = \begin{bmatrix} 6 & 4 \\ 2 & 2 \end{bmatrix} \begin{bmatrix} x \\ y \end{bmatrix},$$

or for the general case:

$$\begin{bmatrix} r \\ s \end{bmatrix} = \begin{bmatrix} a & b \\ c & d \end{bmatrix} \begin{bmatrix} x \\ y \end{bmatrix}.$$

In either case, the original equations are reconstructed using the following rules:

1. Select $r$ followed by '=' and multiply the elements of the top row of the coefficient matrix by the elements of the $x$–$y$ vector respectively:

$$r = ax + by.$$

2. Select $s$ followed by '=' and multiply the elements of the bottom row of the coefficient matrix by the elements of the $x$–$y$ vector respectively:

$$s = cx + dy.$$

For example, the following matrix equation

$$\begin{bmatrix} 30 \\ 20 \end{bmatrix} = \begin{bmatrix} 2 & 3 \\ 4 & 5 \end{bmatrix} \begin{bmatrix} x \\ y \end{bmatrix},$$

represents the pair of linear equations

$$30 = 2x + 3y$$
$$20 = 4x + 5y.$$

The power of matrix notation is that there is no limit to the number of linear equations and variables one can manipulate. For example, the following matrix equation

$$\begin{bmatrix} 10 \\ 20 \\ 30 \end{bmatrix} = \begin{bmatrix} 1 & 2 & 3 \\ 4 & 5 & 6 \\ 7 & 8 & 9 \end{bmatrix} \begin{bmatrix} x \\ y \\ z \end{bmatrix},$$

represents the three linear equations

$$10 = x + 2y + 3z$$
$$20 = 4x + 5y + 6z$$
$$30 = 7x + 8y + 9z.$$

Any pair of linear equations can be represented as a matrix equation, and permits us to write the solution in matrix form. For instance, the solution to the original equations requires the following matrix equation:

$$\begin{bmatrix} x \\ y \end{bmatrix} = \begin{bmatrix} e & f \\ g & h \end{bmatrix} \begin{bmatrix} r \\ s \end{bmatrix},$$

which is another way of writing

$$x = er + fs$$
$$y = gr + hs.$$

But we have already computed two such equations:

$$x = \frac{dr - bs}{ad - bc}$$
$$y = \frac{as - cr}{ad - bc}$$

which can be expressed in matrix form. But before we do this, let's substitute $D = ad - bc$ to simplify the equations:

$$x = \frac{1}{D}(dr - bs) \tag{2.13}$$

$$y = \frac{1}{D}(as - cr). \tag{2.14}$$

Now let's express (2.13) and (2.14) as a matrix equation

$$\begin{bmatrix} x \\ y \end{bmatrix} = \frac{1}{D}\begin{bmatrix} d & -b \\ -c & a \end{bmatrix}\begin{bmatrix} r \\ s \end{bmatrix}. \tag{2.15}$$

So we now have a matrix solution for any pair of linear equations with two unknowns. Let's test (2.15).

*Example* Here are two linearly independent equations:

$$24 = 4x + 3y$$

$$11 = x + 2y$$

where

$$a = 4, \quad b = 3, \quad c = 1, \quad d = 2, \quad r = 24, \quad s = 11, \quad D = 5$$

therefore,

$$\begin{bmatrix} x \\ y \end{bmatrix} = \frac{1}{5}\begin{bmatrix} 2 & -3 \\ -1 & 4 \end{bmatrix}\begin{bmatrix} 24 \\ 11 \end{bmatrix}$$

$$x = \frac{1}{5}(2 \times 24 - 3 \times 11) = 3$$

$$y = \frac{1}{5}(-1 \times 24 + 4 \times 11) = 4$$

which is correct.

Although matrix notation provides the same answers as algebra, so far, it does not appear to offer any advantages. However, these will become apparent as we discover more about matrix notation, such as matrix multiplication, which we cover next.

## 2.3 Matrix Multiplication

So far we have seen how to multiply a vector by a matrix. Now let's discover how to multiply one matrix by another. For example, given

$$\mathbf{M} = \begin{bmatrix} a & b \\ c & d \end{bmatrix}, \quad \mathbf{N} = \begin{bmatrix} e & f \\ g & h \end{bmatrix}$$

what is

$$\mathbf{MN} = \begin{bmatrix} a & b \\ c & d \end{bmatrix}\begin{bmatrix} e & f \\ g & h \end{bmatrix}?$$

We can resolve this problem by solving the same problem using algebra. So let's start by declaring two linear equations of the form

$$x'' = ax' + by' \tag{2.16}$$

$$y'' = cx' + dy' \tag{2.17}$$

where

$$x' = ex + fy \tag{2.18}$$

$$y' = gx + hy. \tag{2.19}$$

Next, we substitute (2.18) and (2.19) in (2.16) and (2.17):

$$x'' = a(ex + fy) + b(gx + hy)$$

$$y'' = c(ex + fy) + d(gx + hy)$$

$$x'' = (ae + bg)x + (af + bh)y \tag{2.20}$$

$$y'' = (ce + dg)x + (cf + dh)y. \tag{2.21}$$

This algebraic answer *must* be the same as that given using matrix notation. So let's set up the same scenario using matrices. We begin with

$$\begin{bmatrix} x'' \\ y'' \end{bmatrix} = \begin{bmatrix} a & b \\ c & d \end{bmatrix} \begin{bmatrix} x' \\ y' \end{bmatrix} \tag{2.22}$$

$$\begin{bmatrix} x' \\ y' \end{bmatrix} = \begin{bmatrix} e & f \\ g & h \end{bmatrix} \begin{bmatrix} x \\ y \end{bmatrix}. \tag{2.23}$$

Next, we substitute (2.23) in (2.22)

$$\begin{bmatrix} x'' \\ y'' \end{bmatrix} = \begin{bmatrix} a & b \\ c & d \end{bmatrix} \begin{bmatrix} e & f \\ g & h \end{bmatrix} \begin{bmatrix} x \\ y \end{bmatrix}. \tag{2.24}$$

The matrix form of (2.20) and (2.21) is

$$\begin{bmatrix} x'' \\ y'' \end{bmatrix} = \begin{bmatrix} ae + bg & af + bh \\ ce + dg & cf + dh \end{bmatrix} \begin{bmatrix} x \\ y \end{bmatrix}, \tag{2.25}$$

which means that

$$\mathbf{MN} = \begin{bmatrix} a & b \\ c & d \end{bmatrix} \begin{bmatrix} e & f \\ g & h \end{bmatrix} = \begin{bmatrix} ae + bg & af + bh \\ ce + dg & cf + dh \end{bmatrix}. \tag{2.26}$$

Equation (2.26) shows how the product **MN** must be evaluated.

The terms of the top row of the first matrix: $a$ and $b$, multiply the terms of the first column of the second matrix $e$ and $g$, giving the result $ae + bg$.

**Table 2.1** Subscripts for matrix multiplication

| $(mn)_{ij}$ | = | $m_{ij} \times n_{ij}$ | + | $m_{ij} \times n_{ij}$ |
|---|---|---|---|---|
| $(mn)_{11}$ | = | $m_{11} \times n_{11}$ | + | $m_{12} \times n_{21}$ |
| $(mn)_{12}$ | = | $m_{11} \times n_{12}$ | + | $m_{12} \times n_{22}$ |
| $(mn)_{21}$ | = | $m_{21} \times n_{11}$ | + | $m_{22} \times n_{21}$ |
| $(mn)_{22}$ | = | $m_{21} \times n_{12}$ | + | $m_{22} \times n_{22}$ |

The terms of the top row of the first matrix: $a$ and $b$, multiply the terms of the second column of the second matrix $f$ and $h$, giving the result $af + bh$.

The terms of the bottom row of the first matrix: $c$ and $d$, multiply the terms of the first column of the second matrix $e$ and $g$, giving the result $ce + dg$.

Finally, the terms of the bottom row of the first matrix: $c$ and $d$, multiply the terms of the second column of the second matrix $f$ and $h$, giving the result $cf + dh$.

Observe that the product result is placed in the common matrix element shared by the row in the first matrix and the column in the second matrix.

To formalise this operation, let's reference any matrix element using the subscripts $(ij)$ where $i$ is the row, and $j$ the column.

Let $m_{ij}$ be an element in **M**, $n_{ij}$ an element in **N**, and $(mn)_{ij}$ be an element in the product **MN**. For example, $m_{11} = a$, $n_{22} = h$ and $(mn)_{12} = af + bh$.

Table 2.1 shows how the four elements of the matrix **MN** are formed from the individual elements of **M** and **N**, which can be generalised to

$$(mn)_{ij} = m_{i1} \times n_{1j} + m_{i2} \times n_{2j}.$$

For example, given

$$\mathbf{M} = \begin{bmatrix} 1 & 2 \\ 3 & 4 \end{bmatrix}, \qquad \mathbf{N} = \begin{bmatrix} 2 & 3 \\ 4 & 5 \end{bmatrix}$$

then

$$\begin{aligned} \mathbf{MN} &= \begin{bmatrix} 1 & 2 \\ 3 & 4 \end{bmatrix} \begin{bmatrix} 2 & 3 \\ 4 & 5 \end{bmatrix} \\ &= \begin{bmatrix} 1 \times 2 + 2 \times 4 & 1 \times 3 + 2 \times 5 \\ 3 \times 2 + 4 \times 4 & 3 \times 3 + 4 \times 5 \end{bmatrix} \\ &= \begin{bmatrix} 10 & 13 \\ 22 & 29 \end{bmatrix}. \end{aligned}$$

Although it may not be immediately obvious, matrix multiplication is non-commutative. i.e. in general, $\mathbf{MN} \neq \mathbf{NM}$. For example,

$$\begin{aligned} \mathbf{NM} &= \begin{bmatrix} e & f \\ g & h \end{bmatrix} \begin{bmatrix} a & b \\ c & d \end{bmatrix} \\ &= \begin{bmatrix} ae + cf & be + df \\ ag + ch & bg + dh \end{bmatrix} \end{aligned}$$

which does not equal **MN** (2.26). Consequently, one has to be careful whenever two or more matrices are multiplied together.

Now that we know how to form the product of two matrices, let's look at a special matrix, which when used to multiply another matrix results in the same matrix.

## 2.4 Identity Matrix

Algebra contains two important objects: the *additive identity* 0 and the *multiplicative identity* 1, which obey the following rules:

$$a + 0 = 0 + a = a$$

$$a \times 1 = 1 \times a = a.$$

The matrix equivalent of 0 is $\mathbf{0}$ which contains only 0s:

$$\mathbf{0} = \begin{bmatrix} 0 & 0 \\ 0 & 0 \end{bmatrix},$$

whereas the matrix equivalent of 1 is not a matrix of 1s, but 0s and 1s. It is easy to discover this matrix, simply by declaring the following pair of linear equations:

$$x = 1x + 0y$$

$$y = 0x + 1y$$

or in matrix form:

$$\begin{bmatrix} x \\ y \end{bmatrix} = \begin{bmatrix} 1 & 0 \\ 0 & 1 \end{bmatrix} \begin{bmatrix} x \\ y \end{bmatrix}. \tag{2.27}$$

The matrix in (2.27) is called an *identity matrix* or a *unit matrix* **I**, and it is easy to see that it has no effect when it pre- or post-multiplies another matrix:

$$\begin{bmatrix} a & b \\ c & d \end{bmatrix} = \begin{bmatrix} 1 & 0 \\ 0 & 1 \end{bmatrix} \begin{bmatrix} a & b \\ c & d \end{bmatrix}$$

$$\begin{bmatrix} a & b \\ c & d \end{bmatrix} = \begin{bmatrix} a & b \\ c & d \end{bmatrix} \begin{bmatrix} 1 & 0 \\ 0 & 1 \end{bmatrix}.$$

Now that we have an identity matrix, let's calculate the matrix equivalent of the *multiplicative inverse*.

## 2.5 Inverse Matrix

The *multiplicative inverse* in algebra is an object that obeys the following rule:

$$a \times a^{-1} = a^{-1} \times a = 1.$$

This suggests that for any matrix $\mathbf{A}$, there is an *inverse matrix* $\mathbf{A}^{-1}$ such that

$$\mathbf{A}\mathbf{A}^{-1} = \mathbf{A}^{-1}\mathbf{A} = \mathbf{I},$$

where $\mathbf{I}$ is the identity matrix. Unfortunately, an inverse matrix is not always possible, but for the moment let's assume that this is the case. Thus, if

$$\mathbf{A} = \begin{bmatrix} a & b \\ c & d \end{bmatrix} \quad \text{and} \quad \mathbf{A}^{-1} = \begin{bmatrix} e & f \\ g & h \end{bmatrix},$$

then $\mathbf{A}\mathbf{A}^{-1} = \mathbf{I}$:

$$\mathbf{A}\mathbf{A}^{-1} = \begin{bmatrix} a & b \\ c & d \end{bmatrix} \begin{bmatrix} e & f \\ g & h \end{bmatrix} = \begin{bmatrix} 1 & 0 \\ 0 & 1 \end{bmatrix}. \tag{2.28}$$

We can compute the inverse matrix by expanding (2.28):

$$\begin{bmatrix} ae+bg & af+bh \\ ce+dg & cf+dh \end{bmatrix} = \begin{bmatrix} 1 & 0 \\ 0 & 1 \end{bmatrix}. \tag{2.29}$$

Equating the matrix elements in (2.29) we have

$$ae + bg = 1,$$

where

$$g = \frac{1-ae}{b}. \tag{2.30}$$

Similarly,

$$ce + dg = 0,$$

where

$$g = -\frac{ce}{d}. \tag{2.31}$$

Therefore equating (2.30) and (2.31) we have

$$\frac{1-ae}{b} = -\frac{ce}{d}$$

$$d - ade = -bce$$

$$e = \frac{d}{ad-bc} = \frac{d}{D}$$

where $D = ad - bc$.

Substituting $e$ in (2.31)

$$g = -\frac{ce}{d} = -\frac{c}{d}\frac{d}{D} = -\frac{c}{D}.$$

Equating the remaining matrix elements in (2.29) we have

$$af + bh = 0,$$

where

$$h = -\frac{af}{b}. \tag{2.32}$$

Similarly,

$$cf + dh = 1,$$

where

$$h = \frac{1 - cf}{d}. \tag{2.33}$$

Therefore, equating (2.32) and (2.33) we have

$$-\frac{af}{b} = \frac{1 - cf}{d}$$
$$adf = -b + bcf$$
$$f = -\frac{b}{D}.$$

Substituting $f$ in (2.32)

$$h = -\frac{af}{b} = \frac{a}{b}\frac{b}{D} = \frac{a}{D}.$$

Substituting $e$, $f$, $g$ and $h$ in $\mathbf{A}^{-1}$ we have

$$\mathbf{A}^{-1} = \begin{bmatrix} e & f \\ g & h \end{bmatrix} = \frac{1}{D}\begin{bmatrix} d & -b \\ -c & a \end{bmatrix},$$

which is the same matrix we computed in (2.15). Therefore, given an invertible matrix $\mathbf{A}$:

$$\mathbf{A} = \begin{bmatrix} a & b \\ c & d \end{bmatrix},$$

its inverse $\mathbf{A}^{-1}$ is given by

$$\mathbf{A}^{-1} = \frac{1}{D}\begin{bmatrix} d & -b \\ -c & a \end{bmatrix},$$

where $D = ad - bc$. For example, given

$$\mathbf{A} = \begin{bmatrix} 6 & 4 \\ 3 & 3 \end{bmatrix},$$

then $D = 6$ and

$$\mathbf{A}^{-1} = \frac{1}{6}\begin{bmatrix} 3 & -4 \\ -3 & 6 \end{bmatrix}.$$

The product $\mathbf{AA}^{-1}$ must equal the identity matrix $\mathbf{I}$:

$$\begin{aligned}\mathbf{AA}^{-1} &= \begin{bmatrix} 6 & 4 \\ 3 & 3 \end{bmatrix}\frac{1}{6}\begin{bmatrix} 3 & -4 \\ -3 & 6 \end{bmatrix} \\ &= \frac{1}{6}\begin{bmatrix} 6 & 0 \\ 0 & 6 \end{bmatrix} \\ &= \begin{bmatrix} 1 & 0 \\ 0 & 1 \end{bmatrix}.\end{aligned}$$

The inverse matrix is an extremely useful object and plays an important role in matrix transformations. It can also be used as an algebraic tool for rearranging matrix equations. For example, given the following matrix equation

$$\begin{bmatrix} r \\ s \end{bmatrix} = \begin{bmatrix} a & b \\ c & d \end{bmatrix}\begin{bmatrix} x \\ y \end{bmatrix},$$

if we let

$$\mathbf{v}' = \begin{bmatrix} r \\ s \end{bmatrix}, \qquad \mathbf{M} = \begin{bmatrix} a & b \\ c & d \end{bmatrix}, \qquad \mathbf{v} = \begin{bmatrix} x \\ y \end{bmatrix}$$

then we can write in shorthand

$$\mathbf{v}' = \mathbf{Mv}. \tag{2.34}$$

Multiplying both sides of (2.34) by $\mathbf{M}^{-1}$ we have

$$\begin{aligned}\mathbf{M}^{-1}\mathbf{v}' &= \mathbf{M}^{-1}\mathbf{Mv} \\ &= \mathbf{Iv}\end{aligned}$$

therefore,

$$\begin{aligned}\mathbf{v} &= \mathbf{M}^{-1}\mathbf{v}' \\ \begin{bmatrix} x \\ y \end{bmatrix} &= \frac{1}{D}\begin{bmatrix} d & -b \\ -c & a \end{bmatrix}\begin{bmatrix} r \\ s \end{bmatrix}.\end{aligned}$$

We will employ this strategy later on when we investigate matrix algebra and transformations.

## 2.6 Worked Examples

*Example 1* Solve the linearly independent equations:

$$9 = 3x + 2y$$
$$1 = 4x - y.$$

Using

$$\begin{bmatrix} x \\ y \end{bmatrix} = \frac{1}{D}\begin{bmatrix} d & -b \\ -c & a \end{bmatrix}\begin{bmatrix} r \\ s \end{bmatrix},$$

where

$$r = 9, \quad s = 1, \quad a = 3, \quad b = 2, \quad c = 4, \quad d = -1$$

then $D = -11$, and

$$\begin{aligned}\begin{bmatrix} x \\ y \end{bmatrix} &= -\frac{1}{11}\begin{bmatrix} -1 & -2 \\ -4 & 3 \end{bmatrix}\begin{bmatrix} 9 \\ 1 \end{bmatrix} \\ &= -\frac{1}{11}\begin{bmatrix} -11 \\ -33 \end{bmatrix} \\ &= \begin{bmatrix} 1 \\ 3 \end{bmatrix}.\end{aligned}$$

*Example 2* Solve the linearly independent equations:

$$7 = 3x - y$$
$$0 = -2x - 4y.$$

Using

$$\begin{bmatrix} x \\ y \end{bmatrix} = \frac{1}{D}\begin{bmatrix} d & -b \\ -c & a \end{bmatrix}\begin{bmatrix} r \\ s \end{bmatrix},$$

where

$$r = 7, \quad s = 0, \quad a = 3, \quad b = -1, \quad c = -2, \quad d = -4,$$

then $D = -14$, and

$$\begin{aligned}\begin{bmatrix} x \\ y \end{bmatrix} &= -\frac{1}{14}\begin{bmatrix} -4 & 1 \\ 2 & 3 \end{bmatrix}\begin{bmatrix} 7 \\ 0 \end{bmatrix} \\ &= -\frac{1}{14}\begin{bmatrix} -28 \\ 14 \end{bmatrix}\end{aligned}$$

$$= \begin{bmatrix} 2 \\ -1 \end{bmatrix}.$$

*Example 3* Solve the trivial, linearly independent equations:

$$1 = x$$
$$1 = y.$$

Using

$$\begin{bmatrix} x \\ y \end{bmatrix} = \frac{1}{D} \begin{bmatrix} d & -b \\ -c & a \end{bmatrix} \begin{bmatrix} r \\ s \end{bmatrix},$$

where

$$r = 1, \qquad s = 1, \qquad a = 1, \qquad b = 0, \qquad c = 0, \qquad d = 1,$$

then $D = 1$, and

$$\begin{bmatrix} x \\ y \end{bmatrix} = \frac{1}{1} \begin{bmatrix} 1 & 0 \\ 0 & 1 \end{bmatrix} \begin{bmatrix} 1 \\ 1 \end{bmatrix}$$
$$= \begin{bmatrix} 1 \\ 1 \end{bmatrix}.$$

*Example 4* Solve the following equations:

$$4 = 6x - 4y$$
$$2 = 3x - 2y.$$

Using

$$\begin{bmatrix} x \\ y \end{bmatrix} = \frac{1}{D} \begin{bmatrix} d & -b \\ -c & a \end{bmatrix} \begin{bmatrix} r \\ s \end{bmatrix},$$

where

$$r = 4, \qquad s = 2, \qquad a = 6, \qquad b = -4, \qquad c = 3, \qquad d = -2,$$

then $D = 0$, which confirms that the equations are not linearly independent. The second equation is half the first.

## 2.7 Summary

Hopefully, this chapter has provided a quick introduction to the ideas behind matrix notation, which is just another way of representing a collection of linear equations.

So far, we have only considered two simultaneous linear equations, but there is no limit—a matrix grows in size as the number of equations increases.

We employ special rules when multiplying matrices to ensure that the result agrees with that obtained using algebra. These rules apply to all of the matrices covered in this book.

As we tend to multiply matrices together, rather than add them, we are particularly interested in the identity matrix, which is equivalent to the number 1 in algebra. We are also interested in the inverse matrix, which when used to multiply the original matrix creates the identity matrix.

We have seen that very obvious patterns arise from the coefficients when solving pairs and triples of linear equations. One reoccurring pattern is given the name 'determinant', and is derived from the associated matrix. For a $2 \times 2$ matrix it is the difference of the cross products. Other rules are employed for $3 \times 3$ and $4 \times 4$ matrices.

Before developing a formal description of matrix algebra, we will explore how to compute the determinant for any size matrix.

# Chapter 3
# Determinants

## 3.1 Introduction

In this chapter we investigate how the determinant evolved as a mathematical object and how its scalar value is computed for matrices of different sizes. Once again, the context is simultaneous linear equations, but this time we employ three equations in three unknowns.

The Babylonians were aware of problems involving three unknowns and knew that three scenarios are required to solve such problems. For example, given three types of corn $A$, $B$ and $C$, bundled up such that their individual weight is unknown:

1 bag of $A$, plus 2 bags of $B$, plus 3 bags of $C$ equals 20 measures
3 bags of $A$, plus 1 bag of $B$, plus 2 bags of $C$ equals 17 measures
2 bags of $A$, plus 1 bag of $B$, plus 1 bag of $C$ equals 11 measures

the object is to discover the weights of the three corn types.

The Babylonians were aware that this is a linear problem and the three scenarios can be scaled up and down, added or subtracted. To resolve the problem, the coefficients of $A$, $B$ and $C$ are represented as a table as shown in Table 3.1.

The objective is to change the coefficients such that one scenario contains two zeros. Table 3.2 shows the coefficients when we subtract scenario **2** from $3 \times$ **1**.

Table 3.3 shows the coefficients when we subtract scenario $3 \times$ **3** from $2 \times$ **2**.

Table 3.4 shows the coefficients when we add $5\times$ scenario **2** to **1**.

Scenario **1** in Table 3.4 shows that $12 \times C = 48$, which makes $C = 4$. Substituting this value in scenario **1** in Table 3.3, makes $B = 3$. Finally, substituting values of $B$ and $C$ in scenario **1** of Table 3.1, makes $A = 2$.

The solution to the problem is $A = 2$, $B = 3$, $C = 4$.

This technique is still used to solve simple linear problems in several unknowns. However, it does require a deal of numerical pattern recognition, and accurate calculations. Nevertheless, it demonstrates that the Babylonians knew that the matrix of coefficients held the secret to the problem's solution.

Let us continue and show how the determinant of a matrix helps us understand how to develop a general-purpose solution.

J. Vince, *Matrix Transforms for Computer Games and Animation*,
DOI 10.1007/978-1-4471-4321-5_3, 

**Table 3.1** Coefficients for the three scenarios

| Corn | Scenarios | | |
|---|---|---|---|
| | **1** | **2** | **3** |
| $A$ | 1 | 3 | 2 |
| $B$ | 2 | 1 | 1 |
| $C$ | 3 | 2 | 1 |
| measures | 20 | 17 | 11 |

**Table 3.2** Subtracting scenario **2** from 3 × **1**

| Corn | Scenarios | | |
|---|---|---|---|
| | **1** | **2** | **3** |
| $A$ | 0 | 3 | 2 |
| $B$ | 5 | 1 | 1 |
| $C$ | 7 | 2 | 1 |
| measures | 43 | 17 | 11 |

**Table 3.3** Subtracting scenario 3 × **3** from 2 × **2**

| Corn | Scenarios | | |
|---|---|---|---|
| | **1** | **2** | **3** |
| $A$ | 0 | 0 | 2 |
| $B$ | 5 | −1 | 1 |
| $C$ | 7 | 1 | 1 |
| measures | 43 | 1 | 11 |

**Table 3.4** Adding 5× scenario **2** to **1**

| Corn | Scenarios | | |
|---|---|---|---|
| | **1** | **2** | **3** |
| $A$ | 0 | 0 | 2 |
| $B$ | 0 | −1 | 1 |
| $C$ | 12 | 1 | 1 |
| measures | 48 | 1 | 7 |

## 3.2 Linear Equations in Three Unknowns

So far, we know that given the following matrix equation

$$\begin{bmatrix} r \\ s \end{bmatrix} = \begin{bmatrix} a & b \\ c & d \end{bmatrix} \begin{bmatrix} x \\ y \end{bmatrix},$$

its solution involves the scalar $ad - bc$, which is the determinant of the matrix. What we will now investigate is the determinant of a $3 \times 3$ matrix. We begin by declaring three linear equations in three unknowns:

$$r = ax + by + cz \tag{3.1}$$

$$s = dx + ey + fz \tag{3.2}$$

$$t = gx + hy + iz. \tag{3.3}$$

The objective is to rearrange the equations such that $y$ and $z$ terms are eliminated, leaving $x$. The procedure is as follows: Multiply (3.1) by $e$ and (3.2) by $b$ to create identical $y$ coefficients:

$$er = aex + bey + cez \tag{3.4}$$

$$bs = bdx + bey + bfz. \tag{3.5}$$

Subtract (3.5) from (3.4) to eliminate the $y$ term:

$$er - bs = (ae - bd)x + (ce - bf)z. \tag{3.6}$$

Equation (3.6) only contains an $x$ and $z$ term. Next, we create another equation containing an $x$ and $z$ term by combining (3.1) and (3.3).

Multiply (3.1) by $h$ and (3.3) by $e$ to create identical $y$ coefficients:

$$hs = dhx + ehy + fhz \tag{3.7}$$

$$et = egx + ehy + eiz. \tag{3.8}$$

Subtract (3.8) from (3.7) to eliminate the $y$ term:

$$hs - et = (dh - eg)x + (fh - ei)z. \tag{3.9}$$

Equation (3.9) only contains an $x$ and $z$ term, and can be combined with (3.6) to eliminate the $z$ term.

Multiply (3.6) by $(fh - ei)$ and (3.9) by $(ce - bf)$:

$$(er - bs)(fh - ei) = (ae - bd)(fh - ei)x + (ce - cf)(fh - ei)z \tag{3.10}$$

$$(hs - et)(ce - bf) = (dh - eg)(ce - bf)x + (ce - bf)(fh - ei)z. \tag{3.11}$$

Subtract (3.11) from (3.10) to eliminate the $z$ term:

$$\begin{aligned}
&(er - bs)(fh - ei) - (hs - et)(ce - bf) \\
&\quad = \big[(ae - bd)(fh - ei) - (dh - eg)(ce - bf)\big]x \\
&eir - fhr - bis + chs - cet + bft \\
&\quad = (aei + bfg + cdh - afh - bdi - ceg)x
\end{aligned}$$

$$eir - fhr - bis + chs - cet + bft = Dx$$

therefore,

$$x = \frac{r(ei - fh) - s(bi - ch) + t(bf - ce)}{D}, \tag{3.12}$$

where

$$D = aei + bfg + cdh - afh - bdi - ceg. \tag{3.13}$$

Using a similar technique of elimination, we can show that

$$y = \frac{-r(di - fg) + s(ai - cg) - t(af - cd)}{D} \tag{3.14}$$

$$z = \frac{r(dh - eg) - s(ah - bg) + t(ae - bd)}{D}. \tag{3.15}$$

The first observation to make is that (3.12), (3.14) and (3.15) share a common denominator $D$, which is the determinant of the matrix of coefficients and written:

$$D = \begin{vmatrix} a & b & c \\ d & e & f \\ g & h & i \end{vmatrix} = aei + bfg + cdh - afh - bdi - ceg.$$

The triples $aei$, $bfg$, $cdh$, $afh$, etc., become apparent if we temporarily extend the determinant as follows:

$$\begin{vmatrix} a & b & c & a & b \\ d & e & f & d & e \\ g & h & i & g & h \end{vmatrix},$$

where

$$aei = \begin{vmatrix} \mathbf{a} & b & c & a & b \\ d & \mathbf{e} & f & d & e \\ g & h & \mathbf{i} & g & h \end{vmatrix}, \quad bfg = \begin{vmatrix} a & \mathbf{b} & c & a & b \\ d & e & \mathbf{f} & d & e \\ g & h & i & \mathbf{g} & h \end{vmatrix},$$

$$cdh = \begin{vmatrix} a & b & \mathbf{c} & a & b \\ d & e & f & \mathbf{d} & e \\ g & h & i & g & \mathbf{h} \end{vmatrix}$$

and the negative terms are

$$-ceg = \begin{vmatrix} a & b & \mathbf{c} & a & b \\ d & \mathbf{e} & f & d & e \\ \mathbf{g} & h & i & g & h \end{vmatrix}, \quad -afh = \begin{vmatrix} a & b & c & \mathbf{a} & b \\ d & e & \mathbf{f} & d & e \\ g & \mathbf{h} & i & g & h \end{vmatrix},$$

$$-bdi = \begin{vmatrix} a & b & c & a & \mathbf{b} \\ d & e & f & \mathbf{d} & e \\ g & h & \mathbf{i} & g & h \end{vmatrix}.$$

This expansion is known as *Sarrus's Rule*, after the French mathematician, J.P. Sarrus (1789–1861), and only works for a $3 \times 3$ matrix. For example, this determinant is expanded as follows:

$$\begin{aligned}
&\begin{vmatrix} 2 & 0 & 4 \\ 3 & 1 & 0 \\ 4 & 2 & 2 \end{vmatrix} \\
&\quad = 2 \times 1 \times 2 + 0 \times 0 \times 4 + 4 \times 3 \times 2 - 4 \times 1 \times 4 - 2 \times 0 \times 2 - 0 \times 3 \times 2 \\
&\quad = 4 + 24 - 16 \\
&\quad = 12.
\end{aligned}$$

The second observation concerns the three numerators of (3.12), (3.14) and (3.15):

$$x = \frac{r(ei - fh) - s(bi - ch) + t(bf - ce)}{D} \tag{3.16}$$

$$y = \frac{-r(di - fg) + s(ai - cg) - t(af - cd)}{D} \tag{3.17}$$

$$z = \frac{r(dh - eg) - s(ah - bg) + t(ae - bd)}{D} \tag{3.18}$$

which can be arranged as

$$\begin{aligned}
x &= \frac{rei + bft + csh - cet - rfh - bsi}{D} \\
y &= \frac{asi + rfg + cdt - csg - aft - rdi}{D} \\
z &= \frac{aet + bsg + rdh - reg - ash - bdt}{D}.
\end{aligned}$$

The numerators are the Sarrus expansion of the following determinants:

$$x = \frac{\begin{vmatrix} r & b & c \\ s & e & f \\ t & h & i \end{vmatrix}}{D} \tag{3.19}$$

$$y = \frac{\begin{vmatrix} a & r & c \\ d & s & f \\ g & t & i \end{vmatrix}}{D} \tag{3.20}$$

$$z = \frac{\begin{vmatrix} a & b & r \\ d & e & s \\ g & h & t \end{vmatrix}}{D}. \tag{3.21}$$

Note that the vector $[r \;\; s \;\; t]^T$ replaces the first column for the $x$ value, the second column for the $y$ value, the third column for the $z$ value. We now have a compact way for solving three simultaneous linear equations in three unknowns. For example, let's create three such equations given $x = 1$, $y = 3$ and $z = 5$:

$$\begin{aligned} 20 &= 2x + y + 3z \\ 15 &= 4x + 2y + z \\ 26 &= 3x + y + 4z. \end{aligned}$$

The determinant $D$ is given by

$$\begin{aligned} D &= \begin{vmatrix} 2 & 1 & 3 \\ 4 & 2 & 1 \\ 3 & 1 & 4 \end{vmatrix} \\ &= 2 \times 2 \times 4 + 1 \times 1 \times 3 + 3 \times 4 \times 1 - 3 \times 2 \times 3 - 2 \times 1 \times 1 - 1 \times 4 \times 4 \\ &= 16 + 3 + 12 - 18 - 2 - 16 \\ &= -5 \end{aligned}$$

and

$$\begin{aligned} x &= \frac{\begin{vmatrix} 20 & 1 & 3 \\ 15 & 2 & 1 \\ 26 & 1 & 4 \end{vmatrix}}{-5} = \frac{-5}{-5} = 1 \\ y &= \frac{\begin{vmatrix} 2 & 20 & 3 \\ 4 & 15 & 1 \\ 3 & 26 & 4 \end{vmatrix}}{-5} = \frac{-15}{-5} = 3 \\ z &= \frac{\begin{vmatrix} 2 & 1 & 20 \\ 4 & 2 & 15 \\ 3 & 1 & 26 \end{vmatrix}}{-5} = \frac{-25}{-5} = 5 \end{aligned}$$

which confirm the original values. Sarrus's rule only applies for a $2 \times 2$ and $3 \times 3$ matrix, and another solution is required for larger matrices.

### 3.2.1 The Laplace Expansion

The French mathematician, Pierre Simon Laplace (1749–1827), invented a general method for computing the determinant of any size matrix. To understand this solution, let's return to the three equations (3.16), (3.17), (3.18) for $x$, $y$ and $z$:

$$\begin{aligned} x &= \frac{r(ei - fh) - s(bi - ch) + t(bf - ce)}{D} \\ y &= \frac{-r(di - fg) + s(ai - cg) - t(af - cd)}{D} \end{aligned}$$

$$z = \frac{r(dh - eg) - s(ah - bg) + t(ae - bd)}{D}$$

which can be rearranged as

$$x = \frac{r(ei - fh) - b(si - ft) + c(sh - et)}{D} \tag{3.22}$$

$$y = \frac{a(si - ft) - r(di - fg) + c(dt - gs)}{D} \tag{3.23}$$

$$z = \frac{a(et - hs) - b(dt - gs) + r(dh - ge)}{D}. \tag{3.24}$$

We need only examine (3.22) to see what's going on. The three numerator terms can be expressed as follows:

$$r(ei - fh) = r\begin{vmatrix} e & f \\ h & i \end{vmatrix}$$

$$-b(si - ft) = -b\begin{vmatrix} s & f \\ t & i \end{vmatrix}$$

$$c(sh - et) = c\begin{vmatrix} s & e \\ t & h \end{vmatrix}.$$

The multipliers $r$, $b$ and $c$ form the top row of the determinant in (3.22), but the middle term $b$ has switched signs. The three determinants are constructed from the second and third rows leaving out the column containing the multiplier:

$$\begin{vmatrix} e & f \\ h & i \end{vmatrix} = \begin{vmatrix} r & b & c \\ s & \mathbf{e} & \mathbf{f} \\ t & \mathbf{h} & \mathbf{i} \end{vmatrix}$$

$$\begin{vmatrix} s & f \\ t & i \end{vmatrix} = \begin{vmatrix} r & b & c \\ \mathbf{s} & e & \mathbf{f} \\ \mathbf{t} & h & \mathbf{i} \end{vmatrix}$$

$$\begin{vmatrix} s & e \\ t & h \end{vmatrix} = \begin{vmatrix} r & b & c \\ \mathbf{s} & \mathbf{e} & f \\ \mathbf{t} & \mathbf{h} & i \end{vmatrix}.$$

It should be no surprise that the determinant using Sarrus's technique is identical to the Laplace expansion, and is confirmed by recomputing the above determinant using the expansion:

$$\begin{vmatrix} 2 & 1 & 3 \\ 4 & 2 & 1 \\ 3 & 1 & 4 \end{vmatrix} = 2\begin{vmatrix} 2 & 1 \\ 1 & 4 \end{vmatrix} - 1\begin{vmatrix} 4 & 1 \\ 3 & 4 \end{vmatrix} + 3\begin{vmatrix} 4 & 2 \\ 3 & 1 \end{vmatrix}$$

$$= 2 \times 7 - 1 \times 13 + 3 \times (-2)$$

$$= 14 - 13 - 6$$
$$= -5.$$

Laplace described his expansion in terms of a determinat's *minor determinants*, which with the associated change of sign, are called *cofactors*. The cofactor $c_{ij}$ of an element $a_{ij}$ is the minor determinant that remains after removing from the original determinant the row $i$ and the column $j$.

For example, in (3.25) the minor of $a_{11}$ is identified by removing the first row and the first column; the minor of $a_{12}$ is identified by removing the first row and the second column; and the minor of $a_{13}$ is identified by removing the first row and the third column.

$$|\mathbf{A}| = \begin{vmatrix} a_{11} & a_{12} & a_{13} \\ a_{21} & a_{22} & a_{23} \\ a_{31} & a_{32} & a_{33} \end{vmatrix}. \tag{3.25}$$

The three minor determinants for $a_{11}$, $a_{12}$ and $a_{13}$ are respectively:

$$A_{11} = \begin{vmatrix} a_{22} & a_{23} \\ a_{32} & a_{33} \end{vmatrix}, \qquad A_{12} = \begin{vmatrix} a_{21} & a_{23} \\ a_{31} & a_{33} \end{vmatrix}, \qquad A_{13} = \begin{vmatrix} a_{21} & a_{22} \\ a_{31} & a_{32} \end{vmatrix}$$

whereas, the three cofactors are

$$c_{11} = +a_{11}A_{11}$$
$$c_{12} = -a_{12}A_{12}$$
$$c_{13} = +a_{13}A_{13}.$$

In general, the minor of $a_{ij}$ is denoted $A_{ij}$.

Laplace proposed the following formulae for selecting the cofactor sign:

$$(-1)^{i+j},$$

which generates the pattern for any size matrix

$$\begin{vmatrix} + & - & + & \dots \\ - & + & - & \dots \\ + & - & + & \dots \\ \dots & \dots & \dots & \dots \end{vmatrix}.$$

Although we have chosen the first row to expand the above determinants, any row, or column may be used. For example, computing the above determinant using the third column we get:

$$\begin{vmatrix} 2 & 1 & 3 \\ 4 & 2 & 1 \\ 3 & 1 & 4 \end{vmatrix} = 3\begin{vmatrix} 4 & 2 \\ 3 & 1 \end{vmatrix} - 1\begin{vmatrix} 2 & 1 \\ 3 & 1 \end{vmatrix} + 4\begin{vmatrix} 2 & 1 \\ 4 & 2 \end{vmatrix}$$

$$\begin{aligned}&= 3 \times (-2) - 1 \times (-1) + 4 \times 0\\&= -6 + 1 + 0\\&= -5\end{aligned}$$

or using the second row we get

$$\begin{aligned}\begin{vmatrix} 2 & 1 & 3 \\ 4 & 2 & 1 \\ 3 & 1 & 4 \end{vmatrix} &= -4\begin{vmatrix} 1 & 3 \\ 1 & 4 \end{vmatrix} + 2\begin{vmatrix} 2 & 3 \\ 3 & 4 \end{vmatrix} - 1\begin{vmatrix} 2 & 1 \\ 3 & 1 \end{vmatrix}\\&= -4 \times 1 + 2 \times (-1) - 1 \times (-1)\\&= -4 - 2 + 1\\&= -5.\end{aligned}$$

## 3.3 Linear Equations in Four Unknowns

The largest matrix we encounter in this book is a $4 \times 4$, and Laplace's expansion is also used to compute its determinant. Let's begin with four simultaneous linear equations in four unknowns:

$$\begin{aligned}r &= ax + by + cz + dw\\s &= ex + fy + gz + hw\\t &= ix + jy + kz + lw\\u &= mx + ny + oz + pw\end{aligned}$$

which in matrix notation is

$$\begin{bmatrix} r \\ s \\ t \\ u \end{bmatrix} = \begin{bmatrix} a & b & c & d \\ e & f & g & h \\ i & j & k & l \\ m & n & o & p \end{bmatrix} \begin{bmatrix} x \\ y \\ z \\ w \end{bmatrix}. \tag{3.26}$$

Using the same technique for solving three equations, we can write the solution as

$$x = \frac{\begin{vmatrix} r & b & c & d \\ s & f & g & h \\ t & j & k & l \\ u & n & o & p \end{vmatrix}}{D}, \qquad y = \frac{\begin{vmatrix} a & r & c & d \\ e & s & g & h \\ i & t & k & l \\ m & u & o & p \end{vmatrix}}{D}$$

$$z = \frac{\begin{vmatrix} a & b & r & d \\ e & f & s & h \\ i & j & t & l \\ m & n & u & p \end{vmatrix}}{D}, \qquad w = \frac{\begin{vmatrix} a & b & c & r \\ e & f & g & s \\ i & j & k & t \\ m & n & o & u \end{vmatrix}}{D}$$

where

$$D=\begin{vmatrix} a & b & c & d\\ e & f & g & h\\ i & j & k & l\\ m & n & o & p \end{vmatrix}.$$

Using the Laplace expansion we can write:

$$x=\frac{r\begin{vmatrix} f & g & h\\ j & k & l\\ n & o & p\end{vmatrix}-b\begin{vmatrix} s & g & h\\ t & k & l\\ u & o & p\end{vmatrix}+c\begin{vmatrix} s & f & h\\ t & j & l\\ u & n & p\end{vmatrix}-d\begin{vmatrix} s & f & g\\ t & j & k\\ u & n & o\end{vmatrix}}{D}$$

$$y=\frac{a\begin{vmatrix} s & g & h\\ t & k & l\\ u & o & p\end{vmatrix}-r\begin{vmatrix} e & g & h\\ i & k & l\\ m & o & p\end{vmatrix}+c\begin{vmatrix} e & s & h\\ i & t & l\\ m & u & p\end{vmatrix}-d\begin{vmatrix} e & s & g\\ i & t & k\\ m & u & o\end{vmatrix}}{D}$$

$$z=\frac{a\begin{vmatrix} f & s & h\\ j & t & l\\ n & u & p\end{vmatrix}-b\begin{vmatrix} e & s & h\\ i & t & l\\ m & u & p\end{vmatrix}+r\begin{vmatrix} e & f & h\\ i & j & l\\ m & n & p\end{vmatrix}-d\begin{vmatrix} e & f & s\\ i & j & t\\ m & n & u\end{vmatrix}}{D}$$

$$w=\frac{a\begin{vmatrix} f & g & s\\ j & k & t\\ n & o & u\end{vmatrix}-b\begin{vmatrix} e & g & s\\ i & k & t\\ m & o & u\end{vmatrix}+c\begin{vmatrix} e & f & s\\ i & j & t\\ m & n & u\end{vmatrix}-r\begin{vmatrix} e & f & g\\ i & j & k\\ m & n & o\end{vmatrix}}{D}$$

where

$$D=a\begin{vmatrix} f & g & h\\ j & k & l\\ n & o & p\end{vmatrix}-b\begin{vmatrix} e & g & h\\ i & k & l\\ m & o & p\end{vmatrix}+c\begin{vmatrix} e & f & h\\ i & j & l\\ m & n & p\end{vmatrix}-d\begin{vmatrix} e & f & g\\ i & j & k\\ m & n & o\end{vmatrix}.$$

The $3 \times 3$ determinants can either be expanded using Sarrus's rule or using the Laplace expansion.

## 3.4 Worked Examples

*Example 1*

$$14 = x + 2y + 3z$$
$$12 = 2x + 2y + 2z$$
$$4 = 2x - 2y + 2z.$$

The determinant $D$ is given by

$$\begin{aligned} D &= \begin{vmatrix} 1 & 2 & 3 \\ 2 & 2 & 2 \\ 2 & -2 & 2 \end{vmatrix} \\ &= 1 \times 2 \times 2 + 2 \times 2 \times 2 + 3 \times 2 \times (-2) \\ &\quad - 1 \times 2 \times (-2) - 2 \times 2 \times 2 - 3 \times 2 \times 2 \\ &= 4 + 8 - 12 + 4 - 8 - 12 \\ &= -16 \end{aligned}$$

and

$$x = \frac{\begin{vmatrix} 14 & 2 & 3 \\ 12 & 2 & 2 \\ 4 & -2 & 2 \end{vmatrix}}{-16} = \frac{-16}{-16} = 1$$

$$y = \frac{\begin{vmatrix} 1 & 14 & 3 \\ 2 & 12 & 2 \\ 2 & 4 & 2 \end{vmatrix}}{-16} = \frac{-32}{-16} = 2$$

$$z = \frac{\begin{vmatrix} 1 & 2 & 14 \\ 2 & 2 & 12 \\ 2 & -2 & 4 \end{vmatrix}}{-16} = \frac{-48}{-16} = 3$$

which is correct.

*Example 2* Solve the four equations in four unknowns, where we know the answer in advance. We begin with $x = 1$, $y = 2$, $z = 3$ and $w = 4$ and four equations:

$$\begin{aligned} 20 &= x + 2y + z + 3w \\ 14 &= 2x + y + 2z + w \\ 14 &= x + 3y + z + w \\ 12 &= 3x + y + z + w. \end{aligned}$$

Therefore,

$$\begin{bmatrix} 20 \\ 14 \\ 14 \\ 12 \end{bmatrix} = \begin{bmatrix} 1 & 2 & 1 & 3 \\ 2 & 1 & 2 & 1 \\ 1 & 3 & 1 & 1 \\ 3 & 1 & 1 & 1 \end{bmatrix} \begin{bmatrix} x \\ y \\ z \\ w \end{bmatrix},$$

and for $x$:

$$x = \frac{\begin{vmatrix} 20 & 2 & 1 & 3 \\ 14 & 1 & 2 & 1 \\ 14 & 3 & 1 & 1 \\ 12 & 1 & 1 & 1 \end{vmatrix}}{D},$$

where

$$D = \begin{vmatrix} 1 & 2 & 1 & 3 \\ 2 & 1 & 2 & 1 \\ 1 & 3 & 1 & 1 \\ 3 & 1 & 1 & 1 \end{vmatrix}.$$

Computing $D$ using the Laplace expansion, we have

$$\begin{aligned} D &= \begin{vmatrix} 1 & 2 & 1 \\ 3 & 1 & 1 \\ 1 & 1 & 1 \end{vmatrix} - 2\begin{vmatrix} 2 & 2 & 1 \\ 1 & 1 & 1 \\ 3 & 1 & 1 \end{vmatrix} + \begin{vmatrix} 2 & 1 & 1 \\ 1 & 3 & 1 \\ 3 & 1 & 1 \end{vmatrix} - 3\begin{vmatrix} 2 & 1 & 2 \\ 1 & 3 & 1 \\ 3 & 1 & 1 \end{vmatrix} \\ &= (1+2+3-1-1-6) - 2(2+6+1-3-2-2) \\ &\quad + (6+3+1-9-2-1) - 3(6+3+2-18-2-1) \\ &= -2-4-2+30 \\ &= 22 \end{aligned}$$

therefore,

$$\begin{aligned} x &= \frac{20\begin{vmatrix} 1 & 2 & 1 \\ 3 & 1 & 1 \\ 1 & 1 & 1 \end{vmatrix} - 2\begin{vmatrix} 14 & 2 & 1 \\ 14 & 1 & 1 \\ 12 & 1 & 1 \end{vmatrix} + \begin{vmatrix} 14 & 1 & 1 \\ 14 & 3 & 1 \\ 12 & 1 & 1 \end{vmatrix} - 3\begin{vmatrix} 14 & 1 & 2 \\ 14 & 3 & 1 \\ 12 & 1 & 1 \end{vmatrix}}{22} \\ &= \frac{20(6-8) - 2(52-54) + (68-64) - 3(82-100)}{22} \\ &= \frac{-40+4+4+54}{22} \\ &= 1 \end{aligned}$$

which is correct. Using a similar technique it can be shown that $y = 2$, $z = 3$ and $w = 4$.

## 3.5 Summary

In this chapter we have seen that a matrix possesses a special scalar value called the determinant. This value is also found when solving groups of simultaneous equations algebraically. The determinant of a $2 \times 2$ matrix is the difference between the diagonal coefficients, whereas we must employ the rules invented by Sarrus and Laplace for a $3 \times 3$ matrix, and that of Laplace for larger matrices.

# Chapter 4
# Matrices

## 4.1 Introduction

In Chap. 2 we introduced the basic ideas behind matrices, and in Chap. 3 we saw that there is scalar value, called the determinant, associated with a matrix. In this chapter we return to matrix notation and explore matrix algebra. As matrix algebra is a relatively large subject and includes matrices of complex numbers, this chapter is confined to topics relevant to computer graphics transforms.

Matrix notation was researched by the British mathematician Arthur Cayley (1821–1895) around 1858. Cayley formalised matrix algebra, along with the American mathematicians Benjamin and Charles Pierce. At the start of the 19th century, the German mathematician Carl Gauss (1777–1855) had shown that transforms were not commutative, i.e. $\mathbf{AB} \neq \mathbf{BA}$, and Cayley's matrix notation would help formalise such observations.

## 4.2 Rectangular and Square Matrices

Now that we know the background to matrices, we can define a matrix as a collection of elements organised in rows and columns in the form of a rectangle or a square. Here are two examples:

$$\begin{bmatrix} 2 & 4 & 6 \\ 1 & 2 & 3 \end{bmatrix}, \qquad \begin{bmatrix} 3 & 5 & 7 \\ 2 & 4 & 6 \\ 1 & 2 & 3 \end{bmatrix}.$$

Any element is identified by two indices describing its location in terms of its row and column. For an element $a_{ij}$, $i$ is the *row index* and $j$ the *column index*:

$$\begin{bmatrix} a_{11} & a_{12} & \cdots & a_{1n} \\ a_{21} & a_{22} & \cdots & a_{2n} \\ \vdots & \vdots & \ddots & \vdots \\ a_{m1} & a_{m2} & \cdots & a_{mn} \end{bmatrix}.$$

J. Vince, *Matrix Transforms for Computer Games and Animation*,
DOI 10.1007/978-1-4471-4321-5_4, © Springer-Verlag London 2012

The number of rows and columns determines the *order* of a matrix. For example, the above matrix has $m$ rows and $n$ columns and is called a matrix of *order* $m$ by $n$. A square matrix with $n$ columns and $n$ rows is called a matrix of order $n$. Given two matrices $\mathbf{A}$ and $\mathbf{B}$ with elements $a_{ij}$ and $b_{ij}$ respectively, they are equal if, and only if, they are of the same order, and $a_{ij} = b_{ij}$ for all pairs $(ij)$. For example, in the following matrices, $\mathbf{A} = \mathbf{B}$, but $\mathbf{A} \neq \mathbf{C}$:

$$\mathbf{A} = \begin{bmatrix} 3 & 5 \\ 2 & 4 \end{bmatrix}, \quad \mathbf{B} = \begin{bmatrix} 3 & 5 \\ 2 & 4 \end{bmatrix}, \quad \mathbf{C} = \begin{bmatrix} 2 & 5 \\ 2 & 4 \end{bmatrix}.$$

## 4.3 Matrix Shorthand

In Chap. 2 we noted that matrix multiplication is non-commutative. We now investigate how matrices behave generally, and we begin by declaring a shorthand notation as

$$\mathbf{A} \equiv [a_{ij}],$$

where $i$ and $j$ are natural numbers. This notation reminds us that $\mathbf{A}$ is a collection of elements $a_{ij}$ arranged as a rectangular or square array.

## 4.4 Matrix Addition and Subtraction

Matrix addition or subtraction is rarely performed in computer graphics transforms, however if it is required, the matrices must be of the same order. Therefore, if $\mathbf{A}$ and $\mathbf{B}$ are of the same order:

$$\mathbf{C} = \mathbf{A} \pm \mathbf{B}$$

$$[c_{ij}] = [a_{ij}] \pm [b_{ij}]$$

where $\mathbf{C}$ has the same order as $\mathbf{A}$ and $\mathbf{B}$.

As real number addition is commutative:

$$c_{ij} = a_{ij} \pm b_{ij} = b_{ij} \pm a_{ij},$$

for all pairs of $ij$, it follows that matrix addition is also commutative:

$$\mathbf{A} + \mathbf{B} = \mathbf{B} + \mathbf{A}.$$

It also follows that as real number addition is associative:

$$(a_{ij} + b_{ij}) + c_{ij} = a_{ij} + (b_{ij} + c_{ij}),$$

for all pairs of $ij$, so too is matrix addition:

$$(\mathbf{A} + \mathbf{B}) + \mathbf{C} = \mathbf{A} + (\mathbf{B} + \mathbf{C}).$$

*Example* Given

$$\mathbf{A} = \begin{bmatrix} 1 & 2 \\ 3 & 4 \end{bmatrix}, \qquad \mathbf{B} = \begin{bmatrix} 3 & 4 \\ 5 & 6 \end{bmatrix}$$

then

$$\begin{aligned} \mathbf{A} + \mathbf{B} &= \begin{bmatrix} 1 & 2 \\ 3 & 4 \end{bmatrix} + \begin{bmatrix} 3 & 4 \\ 5 & 6 \end{bmatrix} \\ &= \begin{bmatrix} 4 & 6 \\ 8 & 10 \end{bmatrix} \\ \mathbf{A} - \mathbf{B} &= \begin{bmatrix} 1 & 2 \\ 3 & 4 \end{bmatrix} - \begin{bmatrix} 3 & 4 \\ 5 & 6 \end{bmatrix} \\ &= \begin{bmatrix} -2 & -2 \\ -2 & -2 \end{bmatrix}. \end{aligned}$$

## 4.5 Matrix Scaling

*Matrix scaling* is the action of multiplying each element of a matrix by a scaling factor. For example, matrix $\mathbf{A}$ is scaled by $\lambda$ as follows:

$$\begin{aligned} \lambda\mathbf{A} &= \lambda[a_{ij}] \\ &= [\lambda a_{ij}] \end{aligned}$$

where each element of $\mathbf{A}$ is multiplied by $\lambda$. If $\lambda = 2$ then

$$\begin{aligned} \mathbf{A} &= \begin{bmatrix} 1 & 2 \\ 3 & 4 \end{bmatrix} \\ \lambda\mathbf{A} &= \begin{bmatrix} 2 & 4 \\ 6 & 8 \end{bmatrix}. \end{aligned}$$

It follows that if the elements of a matrix share a common factor, the factor can be placed outside the matrix. For example,

$$\begin{aligned} \mathbf{B} &= \begin{bmatrix} 10 & 20 \\ 30 & 40 \end{bmatrix} \\ &= 10\begin{bmatrix} 1 & 2 \\ 3 & 4 \end{bmatrix}. \end{aligned}$$

*Example 1* Scale matrix $\mathbf{M}$ by $-10$:

$$\mathbf{M} = \begin{bmatrix} -9 & 10 \\ -20 & 4 \end{bmatrix}$$

$$-10\mathbf{M} = \begin{bmatrix} 90 & -100 \\ 200 & -40 \end{bmatrix}.$$

*Example 2* Identify the common factor of matrix **N**:

$$\mathbf{N} = \begin{bmatrix} 4 & 12 \\ 32 & 40 \end{bmatrix}$$
$$= 4\begin{bmatrix} 1 & 3 \\ 8 & 10 \end{bmatrix}.$$

## 4.6 Matrix Multiplication

In Chap. 2 we saw that for matrix multiplication to be consistent with its algebraic equivalent, matrix multiplication must obey certain rules. For instance, given two matrices **A** and **B**:

$$\mathbf{A} = \begin{bmatrix} a_{11} & a_{12} \\ a_{21} & a_{22} \end{bmatrix}, \qquad \mathbf{B} = \begin{bmatrix} b_{11} & b_{12} \\ b_{21} & b_{22} \end{bmatrix}$$

then

$$\mathbf{AB} = \begin{bmatrix} a_{11}b_{11} + a_{12}b_{21} & a_{11}b_{12} + a_{12}b_{22} \\ a_{21}b_{11} + a_{22}b_{21} & a_{21}b_{12} + a_{22}b_{22} \end{bmatrix}.$$

This can be generalised as follows:

Given two matrices **A** and **B**, where **A** is a matrix of order $m \times p$ with elements $a_{ij}$, and **B** is a matrix of order $p \times n$ with elements $b_{ij}$, then $\mathbf{C} = \mathbf{AB}$ is a matrix of order $m \times n$ with elements $c_{ij}$, where

$$c_{ij} = a_{i1}b_{1j} + a_{i2}b_{2j} + a_{i3}b_{3j} + \cdots + a_{in}b_{nj},$$

which can be expressed as

$$c_{ij} = \sum_{k=1}^{p} a_{ik}b_{kj}. \tag{4.1}$$

For example, two matrices of order 3 are multiplied as follows:

$$\mathbf{A} = \begin{bmatrix} 1 & 2 & 3 \\ 4 & 5 & 6 \\ 7 & 8 & 9 \end{bmatrix}, \qquad \mathbf{B} = \begin{bmatrix} 11 & 12 & 13 \\ 14 & 15 & 16 \\ 17 & 18 & 19 \end{bmatrix}.$$

$$\mathbf{C} = \mathbf{AB}$$
$$= \begin{bmatrix} 1 & 2 & 3 \\ 4 & 5 & 6 \\ 7 & 8 & 9 \end{bmatrix} \begin{bmatrix} 11 & 12 & 13 \\ 14 & 15 & 16 \\ 17 & 18 & 19 \end{bmatrix}$$

$$= \begin{bmatrix} 1\times11+2\times14+3\times17 & 1\times12+2\times15+3\times18 & 1\times13+2\times16+3\times19 \\ 4\times11+5\times14+6\times17 & 4\times12+5\times15+6\times18 & 4\times13+5\times16+6\times19 \\ 7\times11+8\times14+9\times17 & 7\times12+8\times15+9\times18 & 7\times13+8\times16+9\times19 \end{bmatrix}$$

$$= \begin{bmatrix} 90 & 144 & 102 \\ 216 & 231 & 246 \\ 342 & 366 & 390 \end{bmatrix}.$$

Reversing the product such that $\mathbf{C} = \mathbf{BA}$ gives

$$c_{ij} = b_{i1}a_{1j} + b_{i2}a_{2j} + b_{i3}a_{3j} + \cdots + b_{in}a_{nj},$$

which, in general, changes every element of $\mathbf{C}$ and is the reason why matrix multiplication is non-commutative.

In order to form the product $\mathbf{AB}$ of two matrices $\mathbf{A}$ and $\mathbf{B}$, the number of columns in $\mathbf{A}$ must equal the number of rows in $\mathbf{B}$. For example, given

$$\mathbf{A} = \begin{bmatrix} 2 & 4 \\ 3 & 1 \\ 4 & 2 \end{bmatrix}, \qquad \mathbf{B} = \begin{bmatrix} 1 & 2 & 3 & 4 \\ 3 & 1 & 2 & 3 \end{bmatrix}.$$

Then

$$\mathbf{AB} = \begin{bmatrix} 2 & 4 \\ 3 & 1 \\ 4 & 2 \end{bmatrix} \begin{bmatrix} 1 & 2 & 3 & 4 \\ 3 & 1 & 2 & 3 \end{bmatrix}$$

$$= \begin{bmatrix} 2\times1+4\times3 & 2\times2+4\times1 & 2\times3+4\times2 & 2\times4+4\times3 \\ 3\times1+1\times3 & 3\times2+1\times1 & 3\times3+1\times2 & 3\times4+1\times3 \\ 4\times1+2\times3 & 4\times2+2\times1 & 4\times3+2\times2 & 4\times4+2\times3 \end{bmatrix}$$

$$= \begin{bmatrix} 14 & 8 & 14 & 20 \\ 6 & 7 & 11 & 15 \\ 10 & 10 & 16 & 22 \end{bmatrix}.$$

However, the product $\mathbf{BA}$ is not possible, because $\mathbf{B}$ has 4 columns, whereas $\mathbf{A}$ has only 3 rows. Consequently, even though one matrix can premultiply another, it does not hold that the reverse is possible. When two matrices satisfy this rule, they are said to be *conformable*.

### 4.6.1 Vector Scalar Product

Vectors are normally represented by a column matrix, which permits them to be premultiplied by some transforming matrix. However, they may also be represented by a row matrix. For example, **a** is a column vector, and **b** a row vector:

$$\mathbf{a} = \begin{bmatrix} x \\ y \\ z \end{bmatrix}, \qquad \mathbf{b} = [x \quad y \quad z].$$

Either type of vector can be *transposed* into the other by using the following notation:

$$\mathbf{b} = \mathbf{a}^{\mathrm{T}} \quad \text{or} \quad \mathbf{a} = \mathbf{b}^{\mathrm{T}}.$$

More will be said about transposing matrices later in this chapter.

In general, given two vectors:

$$\mathbf{a} = \begin{bmatrix} a_x \\ a_y \\ a_z \end{bmatrix}, \qquad \mathbf{b} = \begin{bmatrix} b_x \\ b_y \\ b_z \end{bmatrix}$$

then

$$\mathbf{a} \bullet \mathbf{b} = a_x b_x + a_y b_y + a_z b_z.$$

We can express this product using matrices as follows:

$$\begin{aligned} \mathbf{a}^{\mathrm{T}}\mathbf{b} &= [a_x \quad a_y \quad a_z] \begin{bmatrix} b_x \\ b_y \\ b_z \end{bmatrix} \\ &= [a_x b_x + a_y b_y + a_z b_z]. \end{aligned}$$

i.e. a matrix with 1 row and 1 column, which is just a scalar quantity.

It is also possible to premultiply a row vector by a column vector, e.g. $\mathbf{c} = \mathbf{a}\mathbf{b}^{\mathrm{T}}$:

$$\mathbf{c} = \begin{bmatrix} a_x \\ a_y \\ a_z \end{bmatrix} [b_x \quad b_y \quad b_z].$$

This product is evaluated as follows:

$$\begin{aligned} \mathbf{c} &= \begin{bmatrix} a_x \\ a_y \\ a_z \end{bmatrix} [b_x \quad b_y \quad b_z] \\ &= \begin{bmatrix} a_x b_x & a_x b_y & a_x b_z \\ a_y b_x & a_y b_y & a_y b_z \\ a_z b_x & a_z b_y & a_z b_z \end{bmatrix}. \end{aligned}$$

For example, given

$$\mathbf{a} = \begin{bmatrix} 1 \\ 2 \\ 3 \end{bmatrix}, \qquad \mathbf{b} = \begin{bmatrix} 5 \\ 6 \\ 7 \end{bmatrix}.$$

Then

$$
\begin{aligned}
\mathbf{ab}^{\mathrm{T}} &= \begin{bmatrix} 1 \\ 2 \\ 3 \end{bmatrix} [5 \quad 6 \quad 7] \\
&= \begin{bmatrix} 1\times 5 & 1\times 6 & 1\times 7 \\ 2\times 5 & 2\times 6 & 2\times 7 \\ 3\times 5 & 3\times 6 & 3\times 7 \end{bmatrix} \\
&= \begin{bmatrix} 5 & 6 & 7 \\ 10 & 12 & 14 \\ 15 & 18 & 21 \end{bmatrix}.
\end{aligned}
$$

### *4.6.2 The Vector Product*

Given a pair of vectors

$$
\mathbf{a} = \begin{bmatrix} a_x \\ a_y \\ a_z \end{bmatrix}, \qquad \mathbf{b} = \begin{bmatrix} b_x \\ b_y \\ b_z \end{bmatrix}
$$

their cross product can be encoded as follows:

$$
\mathbf{a} \times \mathbf{b} = \begin{vmatrix} \mathbf{i} & \mathbf{j} & \mathbf{k} \\ a_x & a_y & a_z \\ b_x & b_y & b_z \end{vmatrix},
$$

which when expanded is

$$
\begin{aligned}
\mathbf{a} \times \mathbf{b} &= a_y b_z \mathbf{i} + a_z b_x \mathbf{j} + a_x b_y \mathbf{k} - a_z b_y \mathbf{i} - a_x b_z \mathbf{j} - a_y b_x \mathbf{k} \\
&= (a_y b_z - a_z b_y)\mathbf{i} + (a_z b_x - a_x b_z)\mathbf{j} + (a_x b_y - a_y b_x)\mathbf{k} \\
&= (a_y b_z - a_z b_y)\mathbf{i} - (a_x b_z - a_z b_x)\mathbf{j} + (a_x b_y - a_y b_x)\mathbf{k}.
\end{aligned}
$$

## 4.7 The Zero Matrix

By definition, all the elements of a *zero matrix* equal zero and is represented by **0**. Here are some examples:

$$
[0 \quad 0], \quad [0 \quad 0 \quad 0], \quad \begin{bmatrix} 0 \\ 0 \end{bmatrix}, \quad \begin{bmatrix} 0 \\ 0 \\ 0 \end{bmatrix}, \quad \begin{bmatrix} 0 & 0 & 0 \\ 0 & 0 & 0 \\ 0 & 0 & 0 \end{bmatrix}.
$$

It follows from the rules of matrix addition that $\mathbf{A} + \mathbf{0} = \mathbf{A}$.

## 4.8 The Negative Matrix

By definition, given a matrix **A** with elements $a_{ij}$, its negative form $-\mathbf{A}$ is defined such that

$$-\mathbf{A} = [-a_{ij}].$$

For example, if

$$\mathbf{A} = \begin{bmatrix} 1 & -2 & 3 \\ -4 & 5 & -6 \\ 7 & -8 & 9 \end{bmatrix},$$

then

$$-\mathbf{A} = \begin{bmatrix} -1 & 2 & -3 \\ 4 & -5 & 6 \\ -7 & 8 & -9 \end{bmatrix}.$$

It follows that $\mathbf{A} + (-\mathbf{A}) = \mathbf{0}$, because

$$\mathbf{A} + (-\mathbf{A}) = [a_{ij}] + [-a_{ij}] = [0_{ij}].$$

## 4.9 Diagonal Matrix

A *diagonal matrix* is a square matrix whose elements are zero, apart from its diagonal:

$$\mathbf{A} = \begin{bmatrix} a_{11} & 0 & \cdots & 0 \\ 0 & a_{22} & \cdots & 0 \\ \vdots & \vdots & \ddots & \vdots \\ 0 & 0 & \cdots & a_{nn} \end{bmatrix},$$

consequently, the determinant of a diagonal matrix must be

$$|\mathbf{A}| = a_{11} \times a_{22} \times \cdots \times a_{nn}.$$

Here is a diagonal matrix with its determinant

$$\mathbf{A} = \begin{bmatrix} 2 & 0 & 0 \\ 0 & 3 & 0 \\ 0 & 0 & 4 \end{bmatrix}$$

$$|\mathbf{A}| = 2 \times 3 \times 4 = 24.$$

Now let's consider the product of two diagonal matrices **A** and **B** with the same order. The general product rule is

$$c_{ij} = a_{i1}b_{1j} + a_{i2}b_{2j} + a_{i3}b_{3j} + \cdots + a_{in}b_{nj},$$

for all $ij$ pairs. As the rows of **A** multiply the columns of **B**, the only time there will be a non-zero result, is when the row and column share a common diagonal element. Consequently, the resulting product is also a diagonal matrix. Let's illustrate this using a $2 \times 2$ and a $3 \times 3$ matrix:

$$\mathbf{A} = \begin{bmatrix} a_{11} & 0 \\ 0 & a_{22} \end{bmatrix}, \qquad \mathbf{B} = \begin{bmatrix} b_{11} & 0 \\ 0 & b_{22} \end{bmatrix}$$

$$\mathbf{AB} = \begin{bmatrix} a_{11} & 0 \\ 0 & a_{22} \end{bmatrix} \begin{bmatrix} b_{11} & 0 \\ 0 & b_{22} \end{bmatrix} = \begin{bmatrix} a_{11}b_{11} & 0 \\ 0 & a_{22}b_{22} \end{bmatrix}$$

which is a diagonal matrix. And for a $3 \times 3$ matrix:

$$\mathbf{A} = \begin{bmatrix} a_{11} & 0 & 0 \\ 0 & a_{22} & 0 \\ 0 & 0 & a_{33} \end{bmatrix}, \qquad \mathbf{B} = \begin{bmatrix} b_{11} & 0 & 0 \\ 0 & b_{22} & 0 \\ 0 & 0 & b_{33} \end{bmatrix}$$

$$\mathbf{AB} = \begin{bmatrix} a_{11} & 0 & 0 \\ 0 & a_{22} & 0 \\ 0 & 0 & a_{33} \end{bmatrix} \begin{bmatrix} b_{11} & 0 & 0 \\ 0 & b_{22} & 0 \\ 0 & 0 & b_{33} \end{bmatrix} = \begin{bmatrix} a_{11}b_{11} & 0 & 0 \\ 0 & a_{22}b_{22} & 0 \\ 0 & 0 & a_{33}b_{33} \end{bmatrix}$$

which is another diagonal matrix.

*Example*

$$\mathbf{A} = \begin{bmatrix} 2 & 0 & 0 \\ 0 & 4 & 0 \\ 0 & 0 & 6 \end{bmatrix}, \qquad \mathbf{B} = \begin{bmatrix} 3 & 0 & 0 \\ 0 & 5 & 0 \\ 0 & 0 & 7 \end{bmatrix}$$

$$\mathbf{AB} = \begin{bmatrix} 6 & 0 & 0 \\ 0 & 20 & 0 \\ 0 & 0 & 42 \end{bmatrix}.$$

It should be clear that matrix multiplication of diagonal matrices is commutative. i.e. $\mathbf{AB} = \mathbf{BA}$.

## 4.10 The Identity Matrix

We have already come across the idea of the *identity matrix*, which corresponds to the number 1 in algebra. Now that we have a definition of a diagonal matrix, we can define an identity matrix as a diagonal matrix with its diagonal elements equal to 1. For example:

$$\mathbf{I}_2 = \begin{bmatrix} 1 & 0 \\ 0 & 1 \end{bmatrix}, \qquad \mathbf{I}_3 = \begin{bmatrix} 1 & 0 & 0 \\ 0 & 1 & 0 \\ 0 & 0 & 1 \end{bmatrix}, \qquad \mathbf{I}_4 = \begin{bmatrix} 1 & 0 & 0 & 0 \\ 0 & 1 & 0 & 0 \\ 0 & 0 & 1 & 0 \\ 0 & 0 & 0 & 1 \end{bmatrix}.$$

Thus when we multiply any matrix by $\mathbf{I}_n$ it leaves the matrix unchanged. We can observe its action using a 4th-order matrix:

$$\mathbf{A} = \begin{bmatrix} a_{11} & a_{12} & a_{13} & a_{14} \\ a_{21} & a_{22} & a_{23} & a_{24} \\ a_{31} & a_{32} & a_{33} & a_{34} \\ a_{41} & a_{42} & a_{43} & a_{44} \end{bmatrix}$$

$$\mathbf{I}_4\mathbf{A} = \begin{bmatrix} 1 & 0 & 0 & 0 \\ 0 & 1 & 0 & 0 \\ 0 & 0 & 1 & 0 \\ 0 & 0 & 0 & 1 \end{bmatrix} \begin{bmatrix} a_{11} & a_{12} & a_{13} & a_{14} \\ a_{21} & a_{22} & a_{23} & a_{24} \\ a_{31} & a_{32} & a_{33} & a_{34} \\ a_{41} & a_{42} & a_{43} & a_{44} \end{bmatrix}$$

$$= \begin{bmatrix} a_{11} & a_{12} & a_{13} & a_{14} \\ a_{21} & a_{22} & a_{23} & a_{24} \\ a_{31} & a_{32} & a_{33} & a_{34} \\ a_{41} & a_{42} & a_{43} & a_{44} \end{bmatrix}.$$

It is common to employ $\mathbf{I}$ as the identity matrix, as its order is determined by the associated matrix. Note that $\mathbf{IA} = \mathbf{AI}$.

## 4.11 The Transposed Matrix

Once a matrix has been defined, its symmetry permits one to perform various geometric operations. For example, we could rotate the matrix elements about a central horizontal or vertical axis. However, one useful operation is to interchange rows and columns. Such a matrix is called a *transposed matrix* and is denoted by $\mathbf{A}^{\mathrm{T}}$. Thus, given a matrix $\mathbf{A}$:

$$\mathbf{A} = \begin{bmatrix} a_{11} & a_{12} & a_{13} & a_{14} \\ a_{21} & a_{22} & a_{23} & a_{24} \\ a_{31} & a_{32} & a_{33} & a_{34} \\ a_{41} & a_{42} & a_{43} & a_{44} \end{bmatrix}$$

$$\mathbf{A}^{\mathrm{T}} = \begin{bmatrix} a_{11} & a_{21} & a_{31} & a_{41} \\ a_{12} & a_{22} & a_{32} & a_{42} \\ a_{13} & a_{23} & a_{33} & a_{43} \\ a_{14} & a_{24} & a_{34} & a_{44} \end{bmatrix}.$$

Only the diagonal elements remain unchanged. For example:

$$\mathbf{A} = \begin{bmatrix} 1 & 2 & 3 & 4 \\ 5 & 6 & 7 & 8 \\ 9 & 10 & 11 & 12 \\ 13 & 14 & 15 & 16 \end{bmatrix}$$

$$\mathbf{A}^{\mathrm{T}} = \begin{bmatrix} 1 & 5 & 9 & 13 \\ 2 & 6 & 10 & 14 \\ 3 & 7 & 11 & 15 \\ 4 & 8 & 12 & 16 \end{bmatrix}.$$

Any order matrix can be transposed. For instance, the transpose of an $n \times m$ matrix is an $m \times n$ matrix:

$$\mathbf{A} = \begin{bmatrix} 1 & 2 \\ 3 & 4 \\ 5 & 6 \end{bmatrix}, \qquad \mathbf{A}^{\mathrm{T}} = \begin{bmatrix} 1 & 3 & 5 \\ 2 & 4 & 6 \end{bmatrix}.$$

Furthermore, $(\mathbf{A}^{\mathrm{T}})^{\mathrm{T}} = \mathbf{A}$ and $(\mathbf{A} + \mathbf{B})^{\mathrm{T}} = \mathbf{A}^{\mathrm{T}} + \mathbf{B}^{\mathrm{T}}$. However, it is **not** true that $(\mathbf{AB})^{\mathrm{T}} = \mathbf{A}^{\mathrm{T}}\mathbf{B}^{\mathrm{T}}$. We can see why by evaluating the transpose of the product of two order 2 matrices. Given

$$\mathbf{A} = \begin{bmatrix} a_{11} & a_{12} \\ a_{21} & a_{22} \end{bmatrix}, \qquad \mathbf{B} = \begin{bmatrix} b_{11} & b_{12} \\ b_{21} & b_{22} \end{bmatrix}$$

then

$$\mathbf{A}^{\mathrm{T}} = \begin{bmatrix} a_{11} & a_{21} \\ a_{12} & a_{22} \end{bmatrix}, \qquad \mathbf{B}^{\mathrm{T}} = \begin{bmatrix} b_{11} & b_{21} \\ b_{12} & b_{22} \end{bmatrix}$$

and

$$\mathbf{AB} = \begin{bmatrix} a_{11}b_{11} + a_{12}b_{21} & a_{11}b_{12} + a_{12}b_{22} \\ a_{21}b_{11} + a_{22}b_{21} & a_{21}b_{12} + a_{22}b_{22} \end{bmatrix}.$$

However,

$$(\mathbf{AB})^{\mathrm{T}} = \begin{bmatrix} a_{11}b_{11} + a_{12}b_{21} & a_{21}b_{11} + a_{22}b_{21} \\ a_{11}b_{12} + a_{12}b_{22} & a_{21}b_{12} + a_{22}b_{22} \end{bmatrix},$$

and

$$\begin{aligned} \mathbf{B}^{\mathrm{T}}\mathbf{A}^{\mathrm{T}} &= \begin{bmatrix} b_{11} & b_{21} \\ b_{12} & b_{22} \end{bmatrix} \begin{bmatrix} a_{11} & a_{21} \\ a_{12} & a_{22} \end{bmatrix} \\ &= \begin{bmatrix} a_{11}b_{11} + a_{12}b_{21} & a_{21}b_{11} + a_{22}b_{21} \\ a_{11}b_{12} + a_{12}b_{22} & a_{21}b_{12} + a_{22}b_{22} \end{bmatrix} \\ &= (\mathbf{AB})^{\mathrm{T}}. \end{aligned}$$

Therefore, $(\mathbf{AB})^{\mathrm{T}} = \mathbf{B}^{\mathrm{T}}\mathbf{A}^{\mathrm{T}}$, whereas

$$\begin{aligned}\mathbf{A}^{\mathrm{T}}\mathbf{B}^{\mathrm{T}} &= \begin{bmatrix} a_{11} & a_{21} \\ a_{12} & a_{22} \end{bmatrix} \begin{bmatrix} b_{11} & b_{21} \\ b_{12} & b_{22} \end{bmatrix} \\ &= \begin{bmatrix} a_{11}b_{11} + a_{21}b_{12} & a_{11}b_{21} + a_{21}b_{22} \\ a_{12}b_{11} + a_{22}b_{12} & a_{12}b_{21} + a_{22}b_{22} \end{bmatrix} \\ &\neq (\mathbf{AB})^{\mathrm{T}}.\end{aligned}$$

We can generalise the above as follows:

Suppose $\mathbf{A}$ is a matrix of order $m \times p$ with elements $a_{ij}$, and $\mathbf{B}$ is of order $p \times n$ with elements $b_{ij}$. Then, if $\mathbf{C} = \mathbf{AB}$, it is of order $m \times n$ with elements $c_{ij}$, where

$$c_{ij} = \sum_{k=1}^{p} a_{ik}b_{kj}.$$

Consequently,

$$\begin{aligned}c_{ij}^{\mathrm{T}} = c_{ji} &= \sum_{k=1}^{p} a_{jk}b_{ki} \\ &= \sum_{k=1}^{p} b_{ik}^{\mathrm{T}}a_{kj}^{\mathrm{T}}\end{aligned}$$

therefore,

$$\mathbf{C}^{\mathrm{T}} = \mathbf{B}^{\mathrm{T}}\mathbf{A}^{\mathrm{T}}.$$

Using a similar technique, one can show that $(\mathbf{ABC}\cdots\mathbf{N})^{\mathrm{T}} = \mathbf{N}^{\mathrm{T}}\cdots\mathbf{C}^{\mathrm{T}}\mathbf{B}^{\mathrm{T}}\mathbf{A}^{\mathrm{T}}$.

*Example* If

$$\mathbf{A} = \begin{bmatrix} 2 & 3 \\ 4 & 2 \end{bmatrix}, \qquad \mathbf{B} = \begin{bmatrix} 1 & 2 \\ 5 & 3 \end{bmatrix}$$

then

$$\begin{aligned}\mathbf{AB} &= \begin{bmatrix} 2 & 3 \\ 4 & 2 \end{bmatrix} \begin{bmatrix} 1 & 2 \\ 5 & 3 \end{bmatrix} \\ &= \begin{bmatrix} 17 & 13 \\ 14 & 14 \end{bmatrix} \\ (\mathbf{AB})^{\mathrm{T}} &= \begin{bmatrix} 17 & 14 \\ 13 & 14 \end{bmatrix} \\ &= \mathbf{B}^{\mathrm{T}}\mathbf{A}^{\mathrm{T}}\end{aligned}$$

$$= \begin{bmatrix} 1 & 5 \\ 2 & 3 \end{bmatrix} \begin{bmatrix} 2 & 4 \\ 3 & 2 \end{bmatrix}$$

$$= \begin{bmatrix} 17 & 14 \\ 13 & 14 \end{bmatrix}.$$

## 4.12 Trace

The *trace* of a square matrix $\mathbf{A}$ is defined as the sum of its diagonal elements and written as $\mathrm{Tr}(\mathbf{A})$. For example:

$$\mathbf{A} = \begin{bmatrix} 1 & 2 & 3 & 4 \\ 2 & 3 & 4 & 5 \\ 3 & 4 & 5 & 6 \\ 4 & 5 & 6 & 7 \end{bmatrix}$$

$$\mathrm{Tr}(\mathbf{A}) = 1 + 3 + 5 + 7 = 16.$$

In Chap. 6 we use the trace of a square matrix to reveal the angle of rotation associated with a rotation matrix. And as we will be using the product of two or more rotation transforms we require to establish that

$$\mathrm{Tr}(\mathbf{AB}) = \mathrm{Tr}(\mathbf{BA}),$$

to reassure ourselves that the trace operation is not sensitive to transform order, and is readily proved as follows.

Given two square matrices $\mathbf{A}$ and $\mathbf{B}$:

$$\mathbf{A} = \begin{bmatrix} a_{11} & \cdots & \cdots & a_{1n} \\ \vdots & a_{22} & \ddots & a_{2n} \\ \vdots & \ddots & \ddots & \vdots \\ a_{n1} & \cdots & \cdots & a_{nn} \end{bmatrix}, \qquad \mathbf{B} = \begin{bmatrix} b_{11} & \cdots & \cdots & b_{1n} \\ \vdots & b_{22} & \ddots & b_{2n} \\ \vdots & \ddots & \ddots & \vdots \\ b_{n1} & \cdots & \cdots & b_{nn} \end{bmatrix}$$

then

$$\mathbf{AB} = \begin{bmatrix} a_{11} & \cdots & \cdots & a_{1n} \\ \vdots & a_{22} & \ddots & a_{2n} \\ \vdots & \ddots & \ddots & \vdots \\ a_{n1} & \cdots & \cdots & a_{nn} \end{bmatrix} \begin{bmatrix} b_{11} & \cdots & \cdots & b_{1n} \\ \vdots & b_{22} & \ddots & b_{2n} \\ \vdots & \ddots & \ddots & \vdots \\ b_{n1} & \cdots & \cdots & b_{nn} \end{bmatrix}$$

$$= \begin{bmatrix} a_{11}b_{11} & \cdots & \cdots & a_{1n} \\ \vdots & a_{22}b_{22} & \ddots & a_{2n} \\ \vdots & \ddots & \ddots & \vdots \\ a_{n1} & \cdots & \cdots & a_{nn}b_{nn} \end{bmatrix}$$

and $\mathrm{Tr}(\mathbf{AB}) = a_{11}b_{11} + a_{22}b_{22} + \cdots + a_{nn}b_{nn}$.

Hopefully, it is obvious that reversing the matrix sequence to **BA** only reverses the $a$ and $b$ scalar elements on the diagonal, and therefore does not affect the trace operation.

## 4.13 Symmetric Matrix

It is worth exploring two types of matrices called *symmetric* and *antisymmetric* matrices, as we refer to them in later chapters. A symmetric matrix is a matrix which equals its own transpose:

$$\mathbf{A} = \mathbf{A}^{\mathrm{T}}.$$

For example, the following matrix is symmetric:

$$\mathbf{A} = \begin{bmatrix} 1 & 3 & 4 \\ 3 & 2 & 4 \\ 4 & 4 & 3 \end{bmatrix}.$$

The symmetric part of any square matrix can be isolated as follows. Given a matrix $\mathbf{A}$ and its transpose $\mathbf{A}^{\mathrm{T}}$

$$\mathbf{A} = \begin{bmatrix} a_{11} & a_{12} & \dots & a_{1n} \\ a_{21} & a_{22} & \dots & a_{2n} \\ \vdots & \vdots & \ddots & \vdots \\ a_{n1} & a_{n2} & \dots & a_{nn} \end{bmatrix}, \qquad \mathbf{A}^{\mathrm{T}} = \begin{bmatrix} a_{11} & a_{21} & \dots & a_{n1} \\ a_{12} & a_{22} & \dots & a_{n2} \\ \vdots & \vdots & \ddots & \vdots \\ a_{1n} & a_{2n} & \dots & a_{nn} \end{bmatrix}$$

their sum is

$$\mathbf{A} + \mathbf{A}^{\mathrm{T}} = \begin{bmatrix} 2a_{11} & a_{12}+a_{21} & \dots & a_{1n}+a_{n1} \\ a_{12}+a_{21} & 2a_{22} & \dots & a_{2n}+a_{n2} \\ \vdots & \vdots & \ddots & \vdots \\ a_{1n}+a_{n1} & a_{2n}+a_{n2} & \dots & 2a_{nn} \end{bmatrix}.$$

By inspection, $\mathbf{A} + \mathbf{A}^{\mathrm{T}}$ is symmetric, and dividing by 2 we have

$$\mathbf{S} = \frac{1}{2}(\mathbf{A} + \mathbf{A}^{\mathrm{T}}),$$

which is defined as the symmetric part of $\mathbf{A}$. For example, given

$$\mathbf{A} = \begin{bmatrix} a_{11} & a_{12} & a_{13} \\ a_{21} & a_{22} & a_{23} \\ a_{31} & a_{32} & a_{33} \end{bmatrix}, \qquad \mathbf{A}^{\mathrm{T}} = \begin{bmatrix} a_{11} & a_{21} & a_{31} \\ a_{12} & a_{22} & a_{32} \\ a_{13} & a_{23} & a_{33} \end{bmatrix}$$

then

$$\mathbf{S} = \frac{1}{2}(\mathbf{A} + \mathbf{A}^{\mathrm{T}})$$

$$= \begin{bmatrix} a_{11} & \frac{a_{12}+a_{21}}{2} & \frac{a_{13}+a_{31}}{2} \\ \frac{a_{12}+a_{21}}{2} & a_{22} & \frac{a_{23}+a_{32}}{2} \\ \frac{a_{13}+a_{31}}{2} & \frac{a_{23}+a_{32}}{2} & a_{33} \end{bmatrix}$$

$$= \begin{bmatrix} a_{11} & \frac{s_3}{2} & \frac{s_2}{2} \\ \frac{s_3}{2} & a_{22} & \frac{s_1}{2} \\ \frac{s_2}{2} & \frac{s_1}{2} & a_{33} \end{bmatrix}$$

where

$$s_1 = a_{23} + a_{32}$$

$$s_2 = a_{13} + a_{31}$$

$$s_3 = a_{12} + a_{21}.$$

Using a real example:

$$\mathbf{A} = \begin{bmatrix} 0 & 1 & 4 \\ 3 & 1 & 4 \\ 4 & 2 & 6 \end{bmatrix}, \qquad \mathbf{A}^{\mathrm{T}} = \begin{bmatrix} 0 & 3 & 4 \\ 1 & 1 & 2 \\ 4 & 4 & 6 \end{bmatrix}$$

$$\mathbf{S} = \begin{bmatrix} 0 & 2 & 4 \\ 2 & 1 & 3 \\ 4 & 3 & 6 \end{bmatrix}$$

which equals its own transpose.

## 4.14 Antisymmetric Matrix

An *antisymmetric matrix* is a matrix whose transpose is its own negative:

$$\mathbf{A}^{\mathrm{T}} = -\mathbf{A},$$

and is also known as a *skew symmetric matrix.*

As the elements of $\mathbf{A}$ and $\mathbf{A}^{\mathrm{T}}$ are related by

$$a_{ij} = -a_{ji}.$$

When $k = i = j$:

$$a_{kk} = -a_{kk},$$

which implies that the diagonal elements must be zero. For example, this is an antisymmetric matrix

$$\begin{bmatrix} 0 & 6 & 2 \\ -6 & 0 & -4 \\ -2 & 4 & 0 \end{bmatrix}.$$

In general, we have

$$\mathbf{A} = \begin{bmatrix} a_{11} & a_{12} & \cdots & a_{1n} \\ a_{21} & a_{22} & \cdots & a_{2n} \\ \vdots & \vdots & \ddots & \vdots \\ a_{n1} & a_{n2} & \cdots & a_{nn} \end{bmatrix}, \qquad \mathbf{A}^{\mathrm{T}} = \begin{bmatrix} a_{11} & a_{21} & \cdots & a_{n1} \\ a_{12} & a_{22} & \cdots & a_{n2} \\ \vdots & \vdots & \ddots & \vdots \\ a_{1n} & a_{2n} & \cdots & a_{nn} \end{bmatrix}$$

and their difference is

$$\mathbf{A} - \mathbf{A}^{\mathrm{T}} = \begin{bmatrix} 0 & a_{12} - a_{21} & \cdots & a_{1n} - a_{n1} \\ -(a_{12} - a_{21}) & 0 & \cdots & a_{2n} - a_{n2} \\ \vdots & \vdots & \ddots & \vdots \\ -(a_{1n} - a_{n1}) & -(a_{2n} - a_{n2}) & \cdots & 0 \end{bmatrix}.$$

It is clear that $\mathbf{A} - \mathbf{A}^{\mathrm{T}}$ is antisymmetric, and dividing by 2 we have

$$\mathbf{Q} = \frac{1}{2}\left(\mathbf{A} - \mathbf{A}^{\mathrm{T}}\right).$$

For example:

$$\mathbf{A} = \begin{bmatrix} a_{11} & a_{12} & a_{13} \\ a_{21} & a_{22} & a_{23} \\ a_{31} & a_{32} & a_{33} \end{bmatrix}, \qquad \mathbf{A}^{\mathrm{T}} = \begin{bmatrix} a_{11} & a_{21} & a_{31} \\ a_{12} & a_{22} & a_{32} \\ a_{13} & a_{23} & a_{33} \end{bmatrix}$$

$$\mathbf{Q} = \begin{bmatrix} 0 & \frac{a_{12}-a_{21}}{2} & \frac{a_{13}-a_{31}}{2} \\ \frac{a_{21}-a_{12}}{2} & 0 & \frac{a_{23}-a_{32}}{2} \\ \frac{a_{31}-a_{13}}{2} & \frac{a_{32}-a_{23}}{2} & 0 \end{bmatrix}$$

and if we maintain some symmetry with the subscripts, we have

$$\mathbf{Q} = \begin{bmatrix} 0 & \frac{a_{12}-a_{21}}{2} & -\frac{a_{31}-a_{13}}{2} \\ -\frac{a_{12}-a_{21}}{2} & 0 & \frac{a_{23}-a_{32}}{2} \\ \frac{a_{31}-a_{13}}{2} & -\frac{a_{23}-a_{32}}{2} & 0 \end{bmatrix}$$

$$= \begin{bmatrix} 0 & \frac{q_3}{2} & -\frac{q_2}{2} \\ -\frac{q_3}{2} & 0 & \frac{q_1}{2} \\ \frac{q_2}{2} & -\frac{q_1}{2} & 0 \end{bmatrix}$$

where

$$q_1 = a_{23} - a_{32}$$

$$q_2 = a_{31} - a_{13}$$

$$q_3 = a_{12} - a_{21}.$$

Using a real example:

$$\mathbf{A}=\begin{bmatrix}0&1&4\\3&1&4\\4&2&6\end{bmatrix},\qquad \mathbf{A}^{\mathrm{T}}=\begin{bmatrix}0&3&4\\1&1&2\\4&4&6\end{bmatrix}$$

$$\mathbf{Q}=\begin{bmatrix}0&-1&0\\1&0&1\\0&-1&0\end{bmatrix}.$$

Furthermore, we have already computed

$$\mathbf{S}=\begin{bmatrix}0&2&4\\2&1&3\\4&3&6\end{bmatrix},$$

and

$$\mathbf{S}+\mathbf{Q}=\begin{bmatrix}0&1&4\\3&1&4\\4&2&6\end{bmatrix}=\mathbf{A}.$$

## 4.15 Inverse Matrix

We have already come across the idea of the *inverse matrix* in Chap. 2, where we saw that it is possible for a square matrix $\mathbf{A}$ to have an inverse form $\mathbf{A}^{-1}$ such that the product $\mathbf{A}\mathbf{A}^{-1}=\mathbf{I}$. When we come onto matrix transforms, we will discover that a matrix $\mathbf{A}$ performs a transformation, such as a rotation about an axis, whilst its inverse $\mathbf{A}^{-1}$ performs the inverse transformation, which rotates in the opposite direction.

So a useful definition for an inverse matrix is: Let $\mathbf{A}$ be a square matrix of order $n$, and $\mathbf{A}^{-1}$ be another square matrix of order $n$, such that their product $\mathbf{A}\mathbf{A}^{-1}=\mathbf{A}^{-1}\mathbf{A}=\mathbf{I}$. This definition preempts the possibility of matrices that do not have an inverse. For example, the matrix $\mathbf{A}$

$$\mathbf{A}=\begin{bmatrix}1&1\\1&1\end{bmatrix},$$

does not have an inverse as its determinant is zero. Therefore, from now on, when we talk about an inverse matrix, we assume the existence of the inverse form.

One way to derive an inverse matrix employs a cofactor matrix, which is based upon the cofactors associated with any matrix element. This was introduced in Chap. 3.

### 4.15.1 Cofactor Matrix

Although the idea of cofactors has been described in the context of determinants, they can also be applied to matrices. For example, let's start with the following matrix and its cofactor matrix

$$\mathbf{A} = \begin{bmatrix} 0 & 1 & 3 \\ 2 & 1 & 4 \\ 4 & 2 & 6 \end{bmatrix}$$

$$\text{cofactor matrix of } \mathbf{A} = \begin{bmatrix} A_{11} & A_{12} & A_{13} \\ A_{21} & A_{22} & A_{23} \\ A_{31} & A_{32} & A_{33} \end{bmatrix}$$

where

$$A_{11} = +\begin{vmatrix} a_{22} & a_{23} \\ a_{32} & a_{33} \end{vmatrix} = +\begin{vmatrix} 1 & 4 \\ 2 & 6 \end{vmatrix} = -2$$

$$A_{12} = -\begin{vmatrix} a_{21} & a_{23} \\ a_{31} & a_{33} \end{vmatrix} = -\begin{vmatrix} 2 & 4 \\ 4 & 6 \end{vmatrix} = 4$$

$$A_{13} = +\begin{vmatrix} a_{21} & a_{23} \\ a_{31} & a_{33} \end{vmatrix} = +\begin{vmatrix} 2 & 1 \\ 4 & 2 \end{vmatrix} = 0$$

$$A_{21} = -\begin{vmatrix} a_{22} & a_{23} \\ a_{32} & a_{33} \end{vmatrix} = -\begin{vmatrix} 1 & 3 \\ 2 & 6 \end{vmatrix} = 0$$

$$A_{22} = +\begin{vmatrix} a_{11} & a_{13} \\ a_{31} & a_{33} \end{vmatrix} = +\begin{vmatrix} 0 & 3 \\ 4 & 6 \end{vmatrix} = -12$$

$$A_{23} = -\begin{vmatrix} a_{11} & a_{12} \\ a_{31} & a_{32} \end{vmatrix} = -\begin{vmatrix} 0 & 1 \\ 4 & 2 \end{vmatrix} = 4$$

$$A_{31} = +\begin{vmatrix} a_{12} & a_{13} \\ a_{22} & a_{23} \end{vmatrix} = +\begin{vmatrix} 1 & 3 \\ 1 & 4 \end{vmatrix} = 1$$

$$A_{32} = -\begin{vmatrix} a_{11} & a_{13} \\ a_{21} & a_{23} \end{vmatrix} = -\begin{vmatrix} 0 & 3 \\ 2 & 4 \end{vmatrix} = 6$$

$$A_{33} = +\begin{vmatrix} a_{11} & a_{12} \\ a_{21} & a_{22} \end{vmatrix} = +\begin{vmatrix} 0 & 1 \\ 2 & 1 \end{vmatrix} = -2$$

therefore, the cofactor matrix of $\mathbf{A}$ is

$$\begin{bmatrix} -2 & 4 & 0 \\ 0 & -12 & 4 \\ 1 & 6 & -2 \end{bmatrix}.$$

It can be shown that the product of a matrix with the transpose of its cofactor matrix has the following form:

$$\mathbf{A}(\text{cofactor matrix of } \mathbf{A})^{\mathrm{T}} = \begin{bmatrix} |\mathbf{A}| & 0 & \cdots & 0 \\ 0 & |\mathbf{A}| & \cdots & 0 \\ \vdots & \vdots & \ddots & \vdots \\ 0 & 0 & 0 & |\mathbf{A}| \end{bmatrix},$$

and dividing throughout by $|\mathbf{A}|$ we have

$$\frac{\mathbf{A}(\text{cofactor matrix of } \mathbf{A})^{\mathrm{T}}}{|\mathbf{A}|} = \mathbf{I},$$

which implies that

$$\mathbf{A}^{-1} = \frac{(\text{cofactor matrix of } \mathbf{A})^{\mathrm{T}}}{|\mathbf{A}|}.$$

Naturally, this assumes that the inverse actually exists.

Let's find the inverse of the above matrix

$$\mathbf{A} = \begin{bmatrix} 0 & 1 & 3 \\ 2 & 1 & 4 \\ 4 & 2 & 6 \end{bmatrix}$$

$$(\text{cofactor matrix of } \mathbf{A}) = \begin{bmatrix} -2 & 4 & 0 \\ 0 & -12 & 4 \\ 1 & 6 & -2 \end{bmatrix}$$

$$(\text{cofactor matrix of } \mathbf{A})^{\mathrm{T}} = \begin{bmatrix} -2 & 0 & 1 \\ 4 & -12 & 6 \\ 0 & 4 & -2 \end{bmatrix}$$

$$|\mathbf{A}| = 1 \times 4 \times 4 + 3 \times 2 \times 2 - 1 \times 2 \times 6 - 3 \times 1 \times 4 = 4$$

$$\mathbf{A}^{-1} = \frac{1}{4} \begin{bmatrix} -2 & 0 & 1 \\ 4 & -12 & 6 \\ 0 & 4 & -2 \end{bmatrix}.$$

Let's check this result by multiplying $\mathbf{A}$ by $\mathbf{A}^{-1}$ which must equal $\mathbf{I}$:

$$\mathbf{A}\mathbf{A}^{-1} = \begin{bmatrix} 0 & 1 & 3 \\ 2 & 1 & 4 \\ 4 & 2 & 6 \end{bmatrix} \frac{1}{4} \begin{bmatrix} -2 & 0 & 1 \\ 4 & -12 & 6 \\ 0 & 4 & -2 \end{bmatrix}$$

$$= \frac{1}{4} \begin{bmatrix} 4 & 0 & 0 \\ 0 & 4 & 0 \\ 0 & 0 & 4 \end{bmatrix}$$

$$= \begin{bmatrix} 1 & 0 & 0 \\ 0 & 1 & 0 \\ 0 & 0 & 1 \end{bmatrix}.$$

Finally, let's compute the inverse matrix of the following matrix using cofactors:

$$\mathbf{A} = \begin{bmatrix} 2 & 3 \\ 4 & -1 \end{bmatrix}$$

$$(\text{cofactor matrix of } \mathbf{A}) = \begin{bmatrix} -1 & -4 \\ -3 & 2 \end{bmatrix}$$

$$(\text{cofactor matrix of } \mathbf{A})^{\mathrm{T}} = \begin{bmatrix} -1 & -3 \\ -4 & 2 \end{bmatrix}$$

$$|\mathbf{A}| = 2 \times (-1) - 3 \times 4 = -14$$

$$\mathbf{A}^{-1} = \frac{1}{14} \begin{bmatrix} 1 & 3 \\ 4 & -2 \end{bmatrix}.$$

In general, the inverse of a $2 \times 2$ matrix is given by

$$\mathbf{A} = \begin{bmatrix} a_{11} & a_{12} \\ a_{21} & a_{22} \end{bmatrix}$$

$$\mathbf{A}^{-1} = \frac{1}{a_{11}a_{22} - a_{12}a_{21}} \begin{bmatrix} a_{22} & -a_{12} \\ -a_{21} & a_{11} \end{bmatrix}$$

which, for the above matrix is

$$\mathbf{A}^{-1} = \frac{-1}{14} \begin{bmatrix} -1 & -3 \\ -4 & 2 \end{bmatrix} = \frac{1}{14} \begin{bmatrix} 1 & 3 \\ 4 & -2 \end{bmatrix}.$$

## 4.16 Orthogonal Matrix

Many of the matrices used in computer graphics are *orthogonal*, which means that their inverse equals their transpose: $\mathbf{A}^{-1} = \mathbf{A}^{\mathrm{T}}$, and makes inversion extremely easy. For example, a 2D rotation matrix is

$$\mathbf{A} = \begin{bmatrix} \cos\theta & -\sin\theta \\ \sin\theta & \cos\theta \end{bmatrix},$$

and its transpose is

$$\mathbf{A}^{\mathrm{T}} = \begin{bmatrix} \cos\theta & \sin\theta \\ -\sin\theta & \cos\theta \end{bmatrix},$$

and their product is

$$\mathbf{A}\mathbf{A}^{\mathrm{T}} = \begin{bmatrix} \cos\theta & -\sin\theta \\ \sin\theta & \cos\theta \end{bmatrix} \begin{bmatrix} \cos\theta & \sin\theta \\ -\sin\theta & \cos\theta \end{bmatrix}$$
$$= \begin{bmatrix} \cos^2\theta + \sin^2\theta & 0 \\ 0 & \cos^2\theta + \sin^2\theta \end{bmatrix}$$
$$= \begin{bmatrix} 1 & 0 \\ 0 & 1 \end{bmatrix}$$

which confirms that $\mathbf{A}^{\mathrm{T}} = \mathbf{A}^{-1}$.

## 4.17 Worked Examples

*Example 1* Compute the sum, difference and product of the following matrices:

$$\mathbf{A} = \begin{bmatrix} 1 & 2 \\ 3 & 4 \end{bmatrix}, \qquad \mathbf{B} = \begin{bmatrix} 4 & 3 \\ 6 & 5 \end{bmatrix}$$
$$\mathbf{A} + \mathbf{B} = \begin{bmatrix} 1 & 2 \\ 3 & 4 \end{bmatrix} + \begin{bmatrix} 4 & 3 \\ 6 & 5 \end{bmatrix} = \begin{bmatrix} 5 & 5 \\ 9 & 9 \end{bmatrix}$$
$$\mathbf{A} - \mathbf{B} = \begin{bmatrix} 1 & 2 \\ 3 & 4 \end{bmatrix} - \begin{bmatrix} 4 & 3 \\ 6 & 5 \end{bmatrix} = \begin{bmatrix} -3 & -1 \\ -3 & -2 \end{bmatrix}$$
$$\mathbf{A}\mathbf{B} = \begin{bmatrix} 1 & 2 \\ 3 & 4 \end{bmatrix} \begin{bmatrix} 4 & 3 \\ 6 & 5 \end{bmatrix} = \begin{bmatrix} 16 & 13 \\ 36 & 29 \end{bmatrix}.$$

*Example 2* Transpose the following matrices.

$$\mathbf{A} = \begin{bmatrix} 6 & 1 \\ 3 & 2 \end{bmatrix}, \qquad \mathbf{B} = [1 \quad 2 \quad 3], \qquad \mathbf{C} = \begin{bmatrix} 8 \\ 2 \\ 4 \end{bmatrix}$$
$$\mathbf{A}^{\mathrm{T}} = \begin{bmatrix} 6 & 3 \\ 1 & 2 \end{bmatrix}, \qquad \mathbf{B}^{\mathrm{T}} = \begin{bmatrix} 1 \\ 2 \\ 3 \end{bmatrix}, \qquad \mathbf{C}^{\mathrm{T}} = [8 \quad 2 \quad 4].$$

*Example 3* Compute $\mathbf{a}\mathbf{b}^{\mathrm{T}}$, given

$$\mathbf{a} = \begin{bmatrix} 1 \\ 2 \\ 3 \end{bmatrix}, \qquad \mathbf{b} = \begin{bmatrix} 3 \\ 4 \\ 5 \end{bmatrix}$$
$$\mathbf{a}\mathbf{b}^{\mathrm{T}} = \begin{bmatrix} 1 \\ 2 \\ 3 \end{bmatrix} [3 \quad 4 \quad 5]$$

$$= \begin{bmatrix} 1\times 3 & 1\times 4 & 1\times 5 \\ 2\times 3 & 2\times 4 & 2\times 5 \\ 3\times 3 & 3\times 4 & 3\times 5 \end{bmatrix}$$

$$= \begin{bmatrix} 3 & 4 & 5 \\ 6 & 8 & 10 \\ 9 & 12 & 15 \end{bmatrix}.$$

*Example 4* Find the trace of the following matrices.

$$\mathbf{A} = \begin{bmatrix} 2 & 7 \\ 1 & -2 \end{bmatrix}, \qquad \mathbf{B} = \begin{bmatrix} 2 & 7 & 3 \\ 6 & 1 & 2 \\ 1 & 0 & 5 \end{bmatrix}, \qquad \mathbf{C} = \begin{bmatrix} 0 & 4 & 3 \\ 9 & 1 & 5 \\ 8 & 3 & 0 \end{bmatrix}$$

$$\mathrm{Tr}(\mathbf{A}) = 0, \qquad \mathrm{Tr}(\mathbf{B}) = 8, \qquad \mathrm{Tr}(\mathbf{C}) = 1.$$

*Example 5* Compute the symmetric and anti-symmetric parts of $\mathbf{A}$, and their sum.

$$\mathbf{A} = \begin{bmatrix} 2 & 5 & 6 \\ 1 & 1 & 4 \\ 6 & 2 & 6 \end{bmatrix}.$$

Given

$$\mathbf{S} = \frac{1}{2}(\mathbf{A} + \mathbf{A}^{\mathrm{T}}),$$

then

$$\mathbf{A}^{\mathrm{T}} = \begin{bmatrix} 2 & 1 & 6 \\ 5 & 1 & 2 \\ 6 & 4 & 6 \end{bmatrix},$$

therefore, the symmetric part is

$$\mathbf{S} = \frac{1}{2}(\mathbf{A} + \mathbf{A}^{\mathrm{T}})$$

$$= \frac{1}{2}\left(\begin{bmatrix} 2 & 5 & 6 \\ 1 & 1 & 4 \\ 6 & 2 & 6 \end{bmatrix} + \begin{bmatrix} 2 & 1 & 6 \\ 5 & 1 & 2 \\ 6 & 4 & 6 \end{bmatrix}\right)$$

$$= \begin{bmatrix} 2 & 3 & 6 \\ 3 & 1 & 3 \\ 6 & 3 & 6 \end{bmatrix}.$$

The anti-symmetric part is

$$\mathbf{Q} = \frac{1}{2}(\mathbf{A} - \mathbf{A}^{\mathrm{T}})$$

$$= \frac{1}{2}\left(\begin{bmatrix} 2 & 5 & 6 \\ 1 & 1 & 4 \\ 6 & 2 & 6 \end{bmatrix} - \begin{bmatrix} 2 & 1 & 6 \\ 5 & 1 & 2 \\ 6 & 4 & 6 \end{bmatrix}\right)$$

$$= \begin{bmatrix} 0 & 2 & 0 \\ -2 & 0 & 1 \\ 0 & -1 & 0 \end{bmatrix}.$$

Their sum is

$$\mathbf{S} + \mathbf{Q} = \begin{bmatrix} 2 & 3 & 6 \\ 3 & 1 & 3 \\ 6 & 3 & 6 \end{bmatrix} + \begin{bmatrix} 0 & 2 & 0 \\ -2 & 0 & 1 \\ 0 & -1 & 0 \end{bmatrix}$$

$$= \begin{bmatrix} 2 & 5 & 6 \\ 1 & 1 & 4 \\ 6 & 2 & 6 \end{bmatrix}$$

$$= \mathbf{A}.$$

*Example 6* Compute the inverse of $\mathbf{A}$, and show that $\mathbf{AA}^{-1} = \mathbf{I}$.

$$\mathbf{A} = \begin{bmatrix} 0 & 2 & 6 \\ 1 & 0 & 2 \\ 2 & 1 & 3 \end{bmatrix}$$

$$(\text{cofactor matrix of } \mathbf{A}) = \begin{bmatrix} -2 & 1 & 1 \\ 0 & -12 & 4 \\ 4 & 6 & -2 \end{bmatrix}$$

$$(\text{cofactor matrix of } \mathbf{A})^{\mathrm{T}} = \begin{bmatrix} -2 & 0 & 4 \\ 1 & -12 & 6 \\ 1 & 4 & -2 \end{bmatrix}$$

$$|\mathbf{A}| = 2 \times 2 \times 2 + 6 \times 1 \times 1 - 2 \times 1 \times 3 = 8$$

$$\mathbf{A}^{-1} = \frac{1}{8}\begin{bmatrix} -2 & 0 & 4 \\ 1 & -12 & 6 \\ 1 & 4 & -2 \end{bmatrix}$$

$$\mathbf{AA}^{-1} = \begin{bmatrix} 0 & 2 & 6 \\ 1 & 0 & 2 \\ 2 & 1 & 3 \end{bmatrix} \frac{1}{8}\begin{bmatrix} -2 & 0 & 4 \\ 1 & -12 & 6 \\ 1 & 4 & -2 \end{bmatrix}$$

$$= \frac{1}{8}\begin{bmatrix} 8 & 0 & 0 \\ 0 & 8 & 0 \\ 0 & 0 & 8 \end{bmatrix}$$

$$= \begin{bmatrix} 1 & 0 & 0 \\ 0 & 1 & 0 \\ 0 & 0 & 1 \end{bmatrix}.$$

## 4.18 Summary

In this chapter we have extended and formalised matrix notation as described in Chap. 2. Although a matrix can be rectangular or square, computer graphics matrix transforms are square, which simplifies the associated algebra.

We have noted that matrices can be negated, scaled, added, subtracted, multiplied and inverted. However, we will discover in the following chapters that only matrix products and inversions are employed on a regular basis.

We have also seen that two special matrices exist: the zero matrix and the identity matrix, that are equivalent to 0 and 1 used in real number algebra.

By swapping row elements with column elements a matrix is transposed, which in the case of orthogonal matrices, inverts the matrix. This is extremely useful in computer graphics transformations.

In the next chapter we investigate 2D and 3D transforms and discover their matrix form.

# Chapter 5
# 2D Matrix Transforms

## 5.1 Introduction

Cartesian coordinates provide a one-to-one relationship between number and shape, such that when we change a shape's coordinates, we transform its geometry. In computer games and animation, the most widely used transforms include scaling, translation, rotation, shearing and reflection.

Transformations are frequently described using vectors, however, in this introductory chapter, I have focussed on points belonging to a 2D shape. The difference is not that important, as the point $(x, y)$ can be regarded as the point a position vector $[x \quad y]^{\mathrm{T}}$ is locating.

## 5.2 Transforms

A point $P(x, y)$ is transformed into $P'(x', y')$ by the following general transform:

$$x' = ax + by$$
$$y' = cx + dy.$$

By using different values for $a$, $b$, $c$ and $d$ we can scale, shear, reflect or rotate a point about the origin, and is represented by the following matrix transform:

$$\begin{bmatrix} x' \\ y' \end{bmatrix} = \begin{bmatrix} a & b \\ c & d \end{bmatrix} \begin{bmatrix} x \\ y \end{bmatrix}. \tag{5.1}$$

The determinant of a matrix transform controls the change of area that occurs to a shape when its coordinates are transformed. For example, in (5.1) the determinant is $ad - bc$, and if we subject the vertices of a unit-square to this transform, we create the scenario shown in Fig. 5.1.

The vertices of the unit-square are transformed as follows:

$$(0, 0) \Rightarrow (0, 0)$$

J. Vince, *Matrix Transforms for Computer Games and Animation*,
DOI 10.1007/978-1-4471-4321-5_5, 

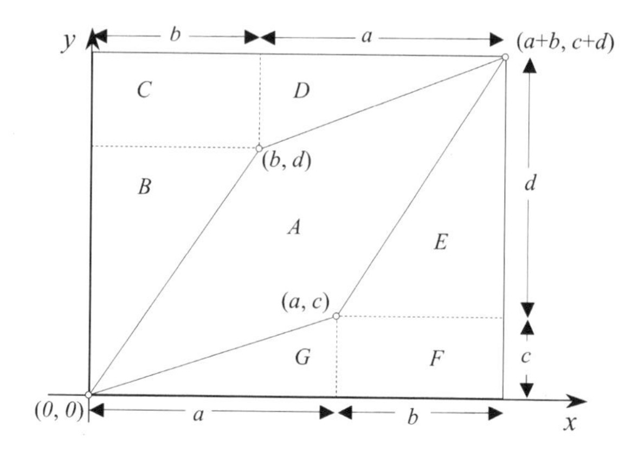


**Fig. 5.1** The inner parallelogram is the transformed unit square

$$
\begin{aligned}
(1, 0) &\Rightarrow (a, c)\\
(1, 1) &\Rightarrow (a + b, c + d)\\
(0, 1) &\Rightarrow (b, d).
\end{aligned}
$$

From Fig. 5.1 it can be seen that the area of the transformed unit-square $A'$ is given by

$$
\begin{aligned}
area &= (a + b)(c + d) - B - C - D - E - F - G\\
&= ac + ad + bc + bd - \frac{1}{2}bd - bc - \frac{1}{2}ac - \frac{1}{2}bd - bc - \frac{1}{2}ac\\
&= ad - bc
\end{aligned}
$$

which is the determinant of the matrix transform. But as the area of the original unit-square is 1, the determinant of the matrix controls the scaling factor applied to the transformed shape.

Returning to (5.1), it may have occurred to you that it cannot effect a translation, as we need to increment both $x$ and $y$ by two offsets $d_x$ and $d_y$:

$$x' = ax + by + d_x \tag{5.2}$$

$$y' = cx + dy + d_y \tag{5.3}$$

which in matrix form is

$$
\begin{bmatrix} x' \\ y' \end{bmatrix} = \begin{bmatrix} a & b \\ c & d \end{bmatrix} \begin{bmatrix} x \\ y \end{bmatrix} + \begin{bmatrix} d_x \\ d_y \end{bmatrix},
$$

and involves a matrix product *and* an addition. Whereas, all the other transforms can be represented by a single matrix product. Fortunately, there is a cunning way around this problem, which entails rewriting (5.2) and (5.3) as

$$x' = ax + by + d_x z \tag{5.4}$$

$$y' = cx + dy + d_y z \tag{5.5}$$

with $z = 1$. The next problem is how to represent (5.4) and (5.5) in matrix form, because the 2D point has effectively been turned into a 3D point with its $z$ coordinate equal to 1? Well, if that is what has happened, let's place the equations in a 3D space by adding a $z$ coordinate with a value of 1:

$$\begin{aligned} x' &= ax + by + d_x \\ y' &= cx + dy + d_y \\ z' &= 0x + 0y + 1 \end{aligned}$$

which has this matrix form:

$$\begin{bmatrix} x' \\ y' \\ z' \end{bmatrix} = \begin{bmatrix} a & b & d_x \\ c & d & d_y \\ 0 & 0 & 1 \end{bmatrix} \begin{bmatrix} x \\ y \\ 1 \end{bmatrix}.$$

Placing the coordinate system into a space with an extra dimension is known as *homogeneous coordinates*, and is widely employed within computer graphics software. As you will see from the following section, the extra coordinate serves a valuable service, and because its value is normally 1, it can be ignored.

### 5.2.1 Homogeneous Coordinates

Homogeneous coordinates surfaced in the early 19th century where they were independently proposed by the German mathematician August Ferdinand Möbius (1790–1868) (who also invented a one-sided curled band, the Möbius strip), Feuerbach, Bobillier, and Plücker. Möbius named them *barycentric coordinates*, and they have also been called *areal coordinates* because of their area-calculating properties.

Homogeneous coordinates define a point in a plane using three coordinates instead of two. Initially, Plücker located a homogeneous point relative to the sides of a triangle, but later revised his notation to the one employed in contemporary mathematics and computer graphics. This states that for a point $(x, y)$ there exists a homogeneous point $(xt, yt, t)$ where $t$ is an arbitrary number. For example, the point $(3, 4)$ has homogeneous coordinates $(6, 8, 2)$, because $3 = 6/2$ and $4 = 8/2$. But the homogeneous point $(6, 8, 2)$ is not unique to $(3, 4)$; $(12, 16, 4)$, $(15, 20, 5)$ and $(300, 400, 100)$ are all possible homogeneous coordinates for $(3, 4)$.

The reason why this coordinate system is called 'homogeneous' is because it is possible to transform functions such as $f(x, y)$ into the form $f(x/t, y/t)$ without disturbing the degree of the curve. To the non-mathematician this may not seem anything to get excited about, but in the field of projective geometry it is a very powerful concept.

For our purposes we can imagine that a collection of homogeneous points of the form $(xt, yt, t)$ exist on an $xy$ plane where $t$ is the $z$ coordinate as illustrated in

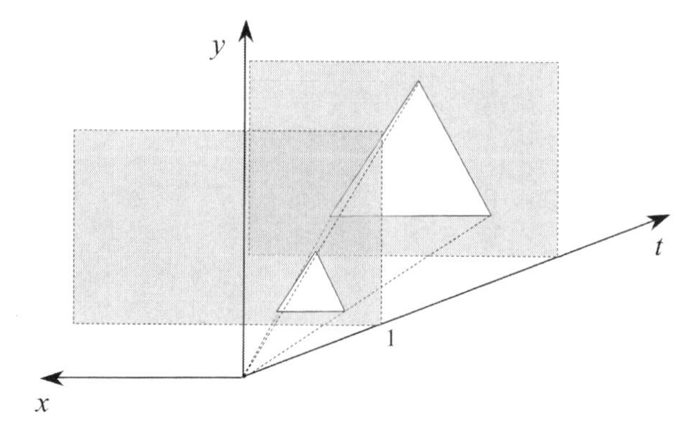


**Fig. 5.2** 2D homogeneous coordinates can be visualised as a plane in 3D space where $t = 1$, for convenience

Fig. 5.2. The figure shows a triangle on the $t = 1$ plane, and a similar triangle, much larger, on a more distant plane. Thus instead of working in two dimensions, we can work on an arbitrary $xy$ plane in three dimensions. The $t$ or $z$ coordinate of the plane is immaterial because the $x$ and $y$ coordinates are eventually scaled by $t$. However, to keep things simple it seems a good idea to choose $t = 1$. This means that the point $(x, y)$ has homogeneous coordinates $(x, y, 1)$ making scaling superfluous.

If we substitute 3D homogeneous coordinates for traditional 2D Cartesian coordinates we must attach 1 to every $(x, y)$ pair. When a point $(x, y, 1)$ is transformed, it emerges as $(x', y', 1)$, and we discard the 1. This may seem a futile exercise, but it resolves the problem of creating a translation transform.

## 5.3 Translation

A translation is represented algebraically by

$$\begin{aligned} x' &= x + d_x \\ y' &= y + d_y \end{aligned}$$

or as a homogeneous matrix transform by

$$\begin{bmatrix} x' \\ y' \\ 1 \end{bmatrix} = \begin{bmatrix} 1 & 0 & d_x \\ 0 & 1 & d_y \\ 0 & 0 & 1 \end{bmatrix} \begin{bmatrix} x \\ y \\ 1 \end{bmatrix}.$$

For example, to translate a shape by $(3, 1)$, as shown in Fig. 5.3, $d_x = 3$ and $d_y = 1$:

$$\begin{bmatrix} x' \\ y' \\ 1 \end{bmatrix} = \begin{bmatrix} 1 & 0 & 3 \\ 0 & 1 & 1 \\ 0 & 0 & 1 \end{bmatrix} \begin{bmatrix} x \\ y \\ 1 \end{bmatrix}.$$

As the transform is applied to every vertex of a shape, there can be no change in the shape's geometry. It is neither rotated nor scaled, and consequently, the shape's

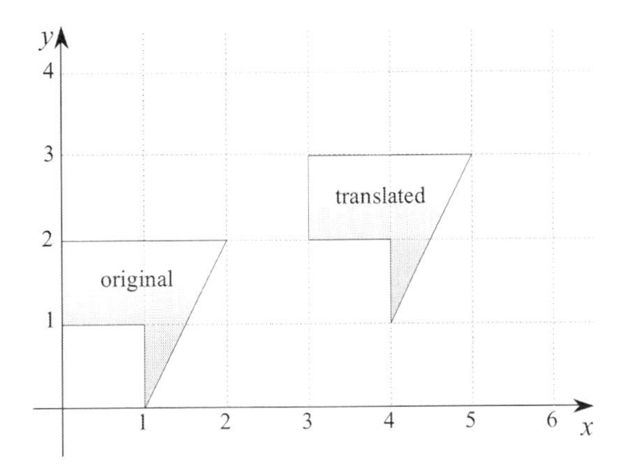


**Fig. 5.3** The shape has been translated by (3, 1)

area is unchanged. Furthermore, because the transform's determinant is a measure of the area change that occurs during the transform, the determinant of a translation matrix must equal 1:

$$\begin{vmatrix} 1 & 0 & d_x \\ 0 & 1 & d_y \\ 0 & 0 & 1 \end{vmatrix} = 1.$$

In Chap. 4 we discovered that for orthogonal matrices, the inverse matrix is its transpose. So, is a translation matrix orthogonal? Well the answer is no, and we can demonstrate this by forming the product of the translation matrix with its transpose to see if it equals the identity matrix:

$$\begin{bmatrix} 1 & 0 & d_x \\ 0 & 1 & d_y \\ 0 & 0 & 1 \end{bmatrix} \begin{bmatrix} 1 & 0 & 0 \\ 0 & 1 & 0 \\ d_x & d_y & 1 \end{bmatrix} = \begin{bmatrix} 1+d_x^2 & d_x d_y & d_x \\ d_x d_y & 1+d_y^2 & d_y \\ d_x & d_y & 1 \end{bmatrix},$$

which is nothing like the identity matrix, therefore a translation matrix is not orthogonal.

We also discovered in Chap. 4 that the inverse matrix is given by

$$\mathbf{T}^{-1} = \frac{(\text{cofactor matrix of } \mathbf{T})^{\mathrm{T}}}{|\mathbf{T}|},$$

and as the determinant equals 1, the inverse translation matrix is

$$\begin{aligned} \mathbf{T}^{-1} &= \begin{bmatrix} 1 & 0 & 0 \\ 0 & 1 & 0 \\ -d_x & -d_y & 1 \end{bmatrix}^{\mathrm{T}} \\ &= \begin{bmatrix} 1 & 0 & -d_x \\ 0 & 1 & -d_y \\ 0 & 0 & 1 \end{bmatrix} \end{aligned}$$

which should be no surprise. One last test confirms that

$$\begin{bmatrix} 1 & 0 & -d_x \\ 0 & 1 & -d_y \\ 0 & 0 & 1 \end{bmatrix}\begin{bmatrix} 1 & 0 & d_x \\ 0 & 1 & d_y \\ 0 & 0 & 1 \end{bmatrix} = \begin{bmatrix} 1 & 0 & 0 \\ 0 & 1 & 0 \\ 0 & 0 & 1 \end{bmatrix}.$$

## 5.4 Scaling

Scaling is represented algebraically by

$$x' = sx$$
$$y' = sy$$

or as a matrix transform by

$$\begin{bmatrix} x' \\ y' \end{bmatrix} = \begin{bmatrix} s & 0 \\ 0 & s \end{bmatrix}\begin{bmatrix} x \\ y \end{bmatrix}.$$

The scaling action is relative to the origin, i.e. the point $(0, 0)$ remains unchanged. All other points move away from the origin when $s > 1$, or move towards the origin when $s < 1$. It is also useful to employ independent scaling factors as follows:

$$\begin{bmatrix} x' \\ y' \end{bmatrix} = \begin{bmatrix} s_x & 0 \\ 0 & s_y \end{bmatrix}\begin{bmatrix} x \\ y \end{bmatrix}.$$

The determinant of a general scaling matrix is given by

$$\begin{vmatrix} s_x & 0 \\ 0 & s_y \end{vmatrix} = s_x s_y,$$

and when $s_x = 2$ and $s_y = 1.5$, the determinant equals 3, which is the area increase observed in Fig. 5.4.

To scale relative to another point $(p_x, p_y)$, we first subtract $(p_x, p_y)$ from $(x, y)$ respectively, which effectively makes the reference point $(p_x, p_y)$ the new origin. Second, we perform the scaling operation relative to the new origin, and third, add $(p_x, p_y)$ back to the new $(x, y)$ respectively, to compensate for the original subtraction. Algebraically this is

$$x' = s_x(x - p_x) + p_x$$
$$y' = s_y(y - p_y) + p_y$$

which simplifies to

$$x' = s_x x + p_x(1 - s_x)$$

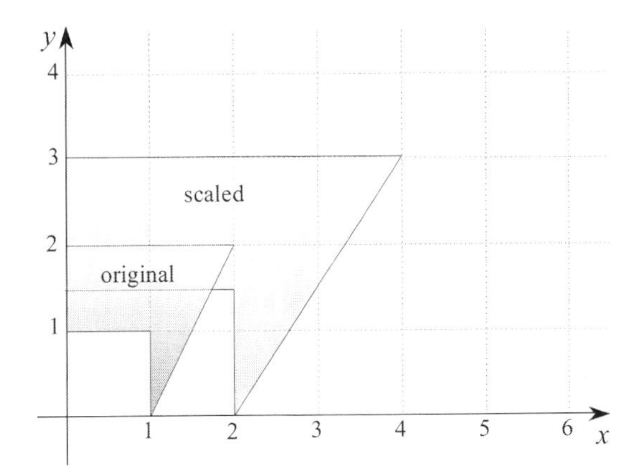


**Fig. 5.4** The scaled shape results by multiplying the $x$ coordinates by 2 and the $y$ coordinates by 1.5

$$y' = s_y y + p_y(1 - s_y)$$

or as a homogeneous matrix

$$\begin{bmatrix} x' \\ y' \\ 1 \end{bmatrix} = \begin{bmatrix} s_x & 0 & p_x(1-s_x) \\ 0 & s_y & p_y(1-s_y) \\ 0 & 0 & 1 \end{bmatrix} \begin{bmatrix} x \\ y \\ 1 \end{bmatrix}. \tag{5.6}$$

For example, to scale a shape by 2 relative to the point $(1, 1)$ the matrix is

$$\begin{bmatrix} x' \\ y' \\ 1 \end{bmatrix} = \begin{bmatrix} 2 & 0 & -1 \\ 0 & 2 & -1 \\ 0 & 0 & 1 \end{bmatrix} \begin{bmatrix} x \\ y \\ 1 \end{bmatrix}.$$

Let's compute the determinant of (5.6) to ensure that the area change equals $s_x s_y$:

$$\begin{vmatrix} s_x & 0 & p_x(1-s_x) \\ 0 & s_y & p_y(1-s_y) \\ 0 & 0 & 1 \end{vmatrix} = s_x s_y.$$

Now let's explore the inverse scaling matrix. To begin with, the basic scaling matrix is not orthogonal. i.e. the inverse is not its transpose:

$$\mathbf{S} = \begin{bmatrix} s_x & 0 \\ 0 & s_y \end{bmatrix}$$

$$\mathbf{S}^{\mathrm{T}} = \begin{bmatrix} s_x & 0 \\ 0 & s_y \end{bmatrix}$$

$$\mathbf{S}\mathbf{S}^{\mathrm{T}} = \begin{bmatrix} s_x & 0 \\ 0 & s_y \end{bmatrix} \begin{bmatrix} s_x & 0 \\ 0 & s_y \end{bmatrix}$$

$$= \begin{bmatrix} s_x^2 & 0 \\ 0 & s_y^2 \end{bmatrix}$$

which is not the identity matrix. Intuition, suggests that this is the inverse matrix:

$$\mathbf{S}^{-1} = \begin{bmatrix} \frac{1}{s_x} & 0 \\ 0 & \frac{1}{s_y} \end{bmatrix},$$

because

$$\mathbf{S}\mathbf{S}^{-1} = \begin{bmatrix} s_x & 0 \\ 0 & s_y \end{bmatrix} \begin{bmatrix} \frac{1}{s_x} & 0 \\ 0 & \frac{1}{s_y} \end{bmatrix} = \mathbf{I}.$$

We could have derived this using

$$\begin{aligned}
\mathbf{S}^{-1} &= \frac{(\text{cofactor matrix of } \mathbf{S})^{\mathrm{T}}}{|\mathbf{S}|} \\
|\mathbf{S}| &= \begin{vmatrix} s_x & 0 \\ 0 & s_y \end{vmatrix} = s_x s_y \\
\mathbf{S}^{-1} &= \frac{1}{s_x s_y} \begin{bmatrix} s_y & 0 \\ 0 & s_x \end{bmatrix}^{\mathrm{T}} \\
&= \begin{bmatrix} \frac{1}{s_x} & 0 \\ 0 & \frac{1}{s_y} \end{bmatrix}.
\end{aligned}$$

Now let's invert the matrix transform that scales relative to some point, (5.6):

$$\begin{aligned}
\mathbf{S}_P &= \begin{bmatrix} s_x & 0 & p_x(1 - s_x) \\ 0 & s_y & p_y(1 - s_y) \\ 0 & 0 & 1 \end{bmatrix} \\
|\mathbf{S}_P| &= s_x s_y \\
\mathbf{S}_P^{-1} &= \frac{1}{s_x s_y} \begin{bmatrix} s_y & 0 & 0 \\ 0 & s_x & 0 \\ -s_y p_x(1 - s_x) & -s_x p_y(1 - s_y) & s_x s_y \end{bmatrix}^{\mathrm{T}} \\
&= \begin{bmatrix} \frac{1}{s_x} & 0 & p_x(1 - \frac{1}{s_x}) \\ 0 & \frac{1}{s_y} & p_y(1 - \frac{1}{s_y}) \\ 0 & 0 & 1 \end{bmatrix}
\end{aligned}$$

which can be confirmed by forming the product $\mathbf{S}_P\mathbf{S}_P^{-1}$:

$$\mathbf{S}_P\mathbf{S}_P^{-1} = \begin{bmatrix} s_x & 0 & p_x(1 - s_x) \\ 0 & s_y & p_y(1 - s_y) \\ 0 & 0 & 1 \end{bmatrix} \begin{bmatrix} \frac{1}{s_x} & 0 & p_x(1 - \frac{1}{s_x}) \\ 0 & \frac{1}{s_y} & p_y(1 - \frac{1}{s_y}) \\ 0 & 0 & 1 \end{bmatrix} = \mathbf{I}.$$

## 5.5 Reflection

We will investigate three ways of reflecting a point: The first is a reflection about the $x$ or $y$ axis; the second is a reflection about any horizontal or vertical axis; and the third is a reflection about an axis passing through the origin.

### 5.5.1 *Reflection About the x and y Axis*

When a point is reflected about the $x$ axis, the sign of its $y$ coordinate is reversed:

$$\begin{aligned} x' &= x \\ y' &= -y \end{aligned}$$

which in matrix form is

$$\begin{bmatrix} x' \\ y' \end{bmatrix} = \begin{bmatrix} 1 & 0 \\ 0 & -1 \end{bmatrix} \begin{bmatrix} x \\ y \end{bmatrix}.$$

Similarly, to reflect a point about the $y$ axis, the sign of its $x$ coordinate is reversed:

$$\begin{aligned} x' &= -x \\ y' &= y \end{aligned}$$

which in matrix form is

$$\begin{bmatrix} x' \\ y' \end{bmatrix} = \begin{bmatrix} -1 & 0 \\ 0 & 1 \end{bmatrix} \begin{bmatrix} x \\ y \end{bmatrix}.$$

Note that the determinant of the reflection matrix is $-1$, which draws our attention to the reversal of the shape's spatial orientation. If the original vertex sequence is clockwise, the reflected vertex sequence is counter-clockwise.

The reflection matrix is orthogonal because its transpose is its inverse:

$$\begin{aligned} \mathbf{R}_x &= \begin{bmatrix} 1 & 0 \\ 0 & -1 \end{bmatrix} \\ \mathbf{R}_x^{\mathrm{T}} &= \begin{bmatrix} 1 & 0 \\ 0 & -1 \end{bmatrix} \\ \mathbf{R}_x\mathbf{R}_x^{\mathrm{T}} &= \begin{bmatrix} 1 & 0 \\ 0 & -1 \end{bmatrix} \begin{bmatrix} 1 & 0 \\ 0 & -1 \end{bmatrix} \\ &= \begin{bmatrix} 1 & 0 \\ 0 & 1 \end{bmatrix}. \end{aligned}$$

### 5.5.2 *Reflection About a Horizontal or Vertical Axis*

To make a reflection about a horizontal or vertical axis we need to introduce some more algebraic deception. For example, to make a reflection about the horizontal axis $y = a_y$, we first subtract $(0, a_y)$ from the point $(x, y)$ respectively. This effectively translates the point $(x, y)$ to the $x$ axis. Next we perform the reflection by reversing the sign of the modified point's $y$ coordinate. And finally, we add $(0, a_y)$ to the reflected $(x, y)$ respectively, to compensate for the original subtraction. Algebraically, the steps are:

$$\begin{aligned}
x' &= x \\
y_1 &= y - a_y \\
y_2 &= -y_1 \\
y' &= -y_1 + a_y \\
&= -y + a_y + a_y \\
&= -y + 2a_y
\end{aligned}$$

or in matrix form

$$\begin{bmatrix} x' \\ y' \\ 1 \end{bmatrix} = \begin{bmatrix} 1 & 0 & 0 \\ 0 & -1 & 2a_y \\ 0 & 0 & 1 \end{bmatrix} \begin{bmatrix} x \\ y \\ 1 \end{bmatrix}.$$

Similarly, to reflect a shape about a vertical $y$ axis, $x = a_x$ the following transform is required:

$$\begin{aligned}
x' &= -x + 2a_x \\
y' &= y
\end{aligned}$$

or in matrix form

$$\begin{bmatrix} x' \\ y' \\ 1 \end{bmatrix} = \begin{bmatrix} -1 & 0 & 2a_x \\ 0 & 1 & 0 \\ 0 & 0 & 1 \end{bmatrix} \begin{bmatrix} x \\ y \\ 1 \end{bmatrix}.$$

Figure 5.5 shows a shape reflected about the $x = 1$ axis. Now let's compute the inverse matrix transform:

$$\mathbf{R}_{a_y} = \begin{bmatrix} 1 & 0 & 0 \\ 0 & -1 & 2a_y \\ 0 & 0 & 1 \end{bmatrix}$$

$$|\mathbf{R}_{a_y}| = -1$$

$$(\text{cofactor matrix of } \mathbf{R}_{a_y})^{\mathrm{T}} = \begin{bmatrix} -1 & 0 & 0 \\ 0 & 1 & -2a_y \\ 0 & 0 & -1 \end{bmatrix}$$

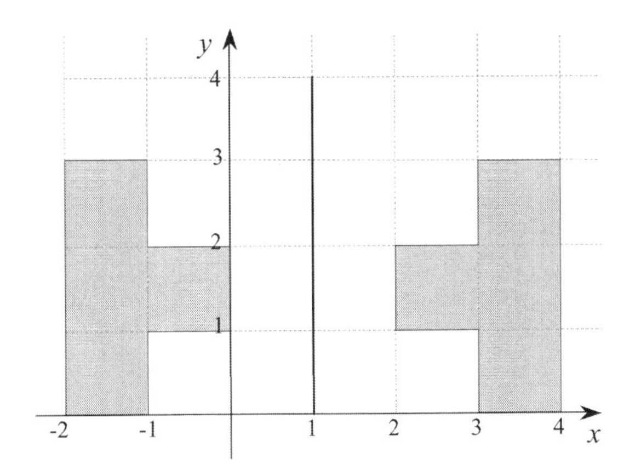


**Fig. 5.5** The shape on the right is reflected about the $x = 1$ axis

$$
\begin{aligned}
\mathbf{R}_{a_y}^{-1} &= \frac{(\text{cofactor matrix of } \mathbf{R}_{a_y})^{\mathrm{T}}}{|\mathbf{R}_{a_y}|} \\
&= \begin{bmatrix} 1 & 0 & 0 \\ 0 & -1 & 2a_y \\ 0 & 0 & 1 \end{bmatrix} = \mathbf{R}_{a_y}.
\end{aligned}
$$

### *5.5.3 Reflection in a Line Intersecting the Origin*

So far we have considered reflections about horizontal and vertical axes, now let's consider a reflection about a line intersecting the origin, as shown in Fig. 5.6. The line of reflection is specified by a unit vector $\mathbf{u}$, and an associated perpendicular vector $\mathbf{n}$. The point to be reflected is $P$ with position vector $\mathbf{p}$, and the reflected point is $P'$ with position vector $\mathbf{p}'$. We define vector $\mathbf{p}$ as

$$\mathbf{p} = a\mathbf{u} + b\mathbf{n},$$

where

$$b = \mathbf{p} \cdot \mathbf{n}.$$

We declare $\mathbf{u}$ and $\mathbf{n}$ as unit column vectors:

$$\mathbf{u} = \begin{bmatrix} u_x \\ u_y \end{bmatrix}, \qquad \mathbf{n} = \begin{bmatrix} -u_y \\ u_x \end{bmatrix}, \qquad u_x^2 + u_y^2 = 1.$$

Therefore,

$$
\begin{aligned}
\mathbf{p}' &= \mathbf{p} - 2b\mathbf{n} \\
&= \mathbf{p} - 2(\mathbf{p} \cdot \mathbf{n})\mathbf{n} \\
&= \mathbf{p} - 2\mathbf{n}(\mathbf{p} \cdot \mathbf{n}).
\end{aligned}
$$

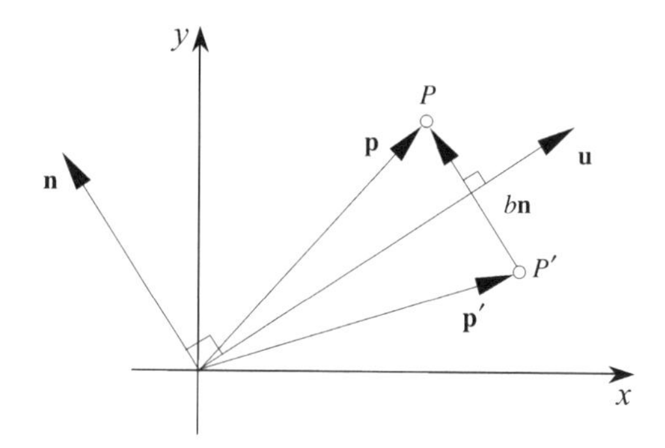


**Fig. 5.6** Reflecting the point $P$ about the vector $\mathbf{u}$ to $P'$

$\mathbf{p}\cdot\mathbf{n}$ is written in matrix form as $\mathbf{n}^{\mathrm{T}}\mathbf{p}$. Therefore,

$$\mathbf{p}' = \mathbf{p} - 2\mathbf{n}\mathbf{n}^{\mathrm{T}}\mathbf{p}.$$

We now have two references to $\mathbf{p}$ and require a mechanism to isolate it. This is achieved by premultiplying $\mathbf{p}$ by the identity matrix $\mathbf{I}$:

$$\begin{aligned}\mathbf{p}' &= \mathbf{I}\mathbf{p} - 2\mathbf{n}\mathbf{n}^{\mathrm{T}}\mathbf{p}\\ &= \left(\mathbf{I} - 2\mathbf{n}\mathbf{n}^{\mathrm{T}}\right)\mathbf{p}\end{aligned}$$

where $\mathbf{I} - 2\mathbf{n}\mathbf{n}^{\mathrm{T}}$ is the transform $\mathbf{R}$:

$$\begin{aligned}\mathbf{R} &= \begin{bmatrix}1 & 0\\ 0 & 1\end{bmatrix} - 2\begin{bmatrix}-u_y\\ u_x\end{bmatrix}[-u_y \quad u_x]\\ &= \begin{bmatrix}1 & 0\\ 0 & 1\end{bmatrix} - 2\begin{bmatrix}u_y^2 & -u_x u_y\\ -u_x u_y & u_x^2\end{bmatrix}\\ &= \begin{bmatrix}1-2u_y^2 & 2u_x u_y\\ 2u_x u_y & 1-2u_x^2\end{bmatrix}\\ &= \begin{bmatrix}u_x^2-u_y^2 & 2u_x u_y\\ 2u_x u_y & u_y^2-u_x^2\end{bmatrix}.\end{aligned}$$

Therefore, we have:

$$\begin{bmatrix}x'\\ y'\end{bmatrix} = \begin{bmatrix}u_x^2-u_y^2 & 2u_x u_y\\ 2u_x u_y & u_y^2-u_x^2\end{bmatrix}\begin{bmatrix}x\\ y\end{bmatrix}.$$

Let's compute the determinant of $\mathbf{R}$ to show that it equals $-1$:

$$\begin{aligned}|\mathbf{R}| &= \begin{vmatrix}u_x^2-u_y^2 & 2u_x u_y\\ 2u_x u_y & u_y^2-u_x^2\end{vmatrix}\\ &= \left(u_x^2-u_y^2\right)\left(u_y^2-u_x^2\right) - (2u_x u_y)(2u_x u_y)\end{aligned}$$

$$
\begin{aligned}
&= u_x^2 u_y^2 - u_y^4 - u_x^4 + u_x^2 u_y^2 - 4u_x^2 u_y^2 \\
&= -\left(u_x^4 + 2u_x^2 u_y^2 + u_y^4\right) \\
&= -\left(u_x^2 + u_y^2\right)^2 \\
&= -1.
\end{aligned}
$$

Now let's test **R** by making **u** intersect the origin at 45°, where the point $P(1, 0)$ is reflected to $P'(0, 1)$:

$$
\begin{aligned}
\mathbf{u}^{\mathrm{T}} &= \begin{bmatrix} \frac{1}{\sqrt{2}} & \frac{1}{\sqrt{2}} \end{bmatrix} \\
P' &= \begin{bmatrix} \frac{1}{2} - \frac{1}{2} & 1 \\ 1 & \frac{1}{2} - \frac{1}{2} \end{bmatrix} \begin{bmatrix} 1 \\ 0 \end{bmatrix} \\
&= \begin{bmatrix} 0 & 1 \\ 1 & 0 \end{bmatrix} \begin{bmatrix} 1 \\ 0 \end{bmatrix} \\
&= \begin{bmatrix} 0 \\ 1 \end{bmatrix}.
\end{aligned}
$$

## 5.6 Shearing

We can shear a shape along the $x$ or $y$ axis by an angle $\alpha$. Figure 5.7 shows the scenario for a shear along the $x$ axis, where we see that the $y$ coordinates remain unchanged but the $x$ coordinates are a function of $y$ and $\tan\alpha$.

$$
\begin{aligned}
x' &= x + y \tan\alpha \\
y' &= y
\end{aligned}
$$

or in matrix form

$$
\begin{bmatrix} x' \\ y' \end{bmatrix} = \begin{bmatrix} 1 & \tan\alpha \\ 0 & 1 \end{bmatrix} \begin{bmatrix} x \\ y \end{bmatrix}.
$$

The determinant equals 1, which confirms that there is no change in area after the shear transform.

As $\tan(-\alpha) = -\tan\alpha$, the direction of shear is determined by the sign of the angle, which makes the inverse transform equal to:

$$
\begin{bmatrix} 1 & -\tan\alpha \\ 0 & 1 \end{bmatrix},
$$

which is confirmed by the transposed cofactor matrix technique.

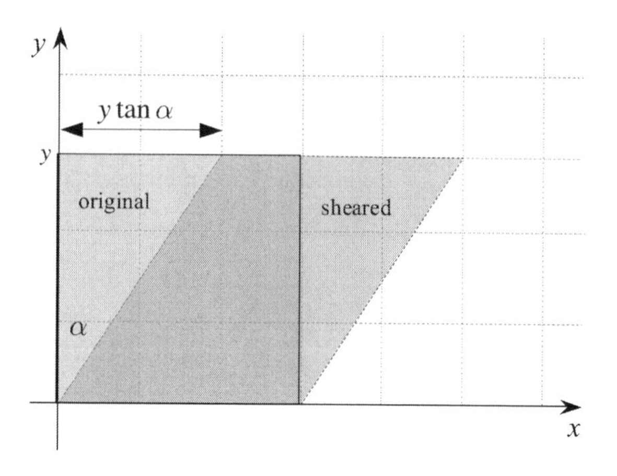


**Fig. 5.7** The shape is sheared along the $x$ axis by angle $\alpha$

Similarly, the transform for shearing along the $y$ axis is given by

$$x' = x$$
$$y' = y + x \tan\alpha$$

or in matrix form

$$\begin{bmatrix} x' \\ y' \end{bmatrix} = \begin{bmatrix} 1 & 0 \\ \tan\alpha & 1 \end{bmatrix} \begin{bmatrix} x \\ y \end{bmatrix}.$$

## 5.7 Rotation

In 2D there are two rotation transforms: the first is about the origin, and the second is about some arbitrary point. We will begin with the former.

Figure 5.8 shows a point $P(x, y)$ which is rotated an angle $\beta$ about the origin to $P'(x', y')$, and as we are dealing with a pure rotation, both $P'$ and $P$ are distance $R$ from the origin.

From Fig. 5.8 it can be seen that

$$\cos\theta = x/R$$
$$\sin\theta = y/R$$
$$x' = R\cos(\theta + \beta)$$
$$y' = R\sin(\theta + \beta)$$

and substituting the identities for $\cos(\theta + \beta)$ and $\sin(\theta + \beta)$ we have

$$x' = R(\cos\theta\cos\beta - \sin\theta\sin\beta)$$
$$y' = R(\sin\theta\cos\beta + \cos\theta\sin\beta)$$

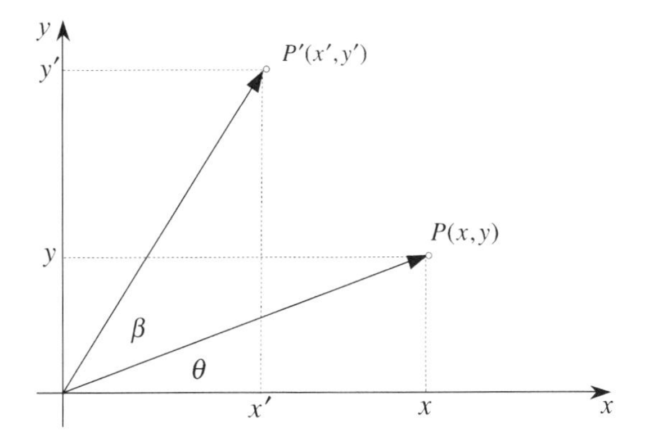


**Fig. 5.8** The point $P(x, y)$ is rotated through an angle $\beta$ to $P'(x', y')$

$$
\begin{aligned}
x' &= R\left(\frac{x}{R}\cos\beta - \frac{y}{R}\sin\beta\right)\\
y' &= R\left(\frac{y}{R}\cos\beta + \frac{x}{R}\sin\beta\right)\\
x' &= x\cos\beta - y\sin\beta\\
y' &= x\sin\beta + y\cos\beta
\end{aligned}
$$

or in matrix form

$$
\begin{bmatrix} x' \\ y' \end{bmatrix} = \begin{bmatrix} \cos\beta & -\sin\beta \\ \sin\beta & \cos\beta \end{bmatrix} \begin{bmatrix} x \\ y \end{bmatrix},
$$

and is the transform for rotating points about the origin.

For example, to rotate a point 90° about the origin the transform becomes

$$
\begin{bmatrix} x' \\ y' \end{bmatrix} = \begin{bmatrix} 0 & -1 \\ 1 & 0 \end{bmatrix} \begin{bmatrix} x \\ y \end{bmatrix}.
$$

Thus the point $(1, 0)$ becomes $(0, 1)$.

Rotating a point 360° about the origin the transform becomes the identity matrix:

$$
\begin{bmatrix} x' \\ y' \end{bmatrix} = \begin{bmatrix} 1 & 0 \\ 0 & 1 \end{bmatrix} \begin{bmatrix} x \\ y \end{bmatrix}.
$$

The following observations can be made about the rotation matrix $\mathbf{R}_\beta$:

$$
\mathbf{R}_\beta = \begin{bmatrix} \cos\beta & -\sin\beta \\ \sin\beta & \cos\beta \end{bmatrix}.
$$

Its determinant equals 1:

$$
|\mathbf{R}_\beta| = \cos^2\beta + \sin^2\beta = 1,
$$

which confirms that there is no change in area when a shape is rotated.

Its transpose is

$$\mathbf{R}_\beta^{\mathrm{T}} = \begin{bmatrix} \cos\beta & \sin\beta \\ -\sin\beta & \cos\beta \end{bmatrix}.$$

The product $\mathbf{R}_\beta \mathbf{R}_\beta^{\mathrm{T}} = \mathbf{I}$:

$$\mathbf{R}_\beta \mathbf{R}_\beta^{\mathrm{T}} = \begin{bmatrix} \cos\beta & -\sin\beta \\ \sin\beta & \cos\beta \end{bmatrix} \begin{bmatrix} \cos\beta & \sin\beta \\ -\sin\beta & \cos\beta \end{bmatrix} = \begin{bmatrix} 1 & 0 \\ 0 & 1 \end{bmatrix},$$

and because $\mathbf{R}_\beta \mathbf{R}_\beta^{\mathrm{T}}$ equals the identity matrix, $\mathbf{R}_\beta^{-1} = \mathbf{R}_\beta^{\mathrm{T}}$:

$$\mathbf{R}_\beta^{-1} = \begin{bmatrix} \cos\beta & \sin\beta \\ -\sin\beta & \cos\beta \end{bmatrix},$$

means that $\mathbf{R}_\beta$ is orthogonal.

### 5.7.1 Rotation About an Arbitrary Point

Now let's see how to rotate a point $(x, y)$ about an arbitrary point $(d_x, d_y)$. The strategy involves making the point of rotation a temporary origin, which is achieved by subtracting $(d_x, d_y)$ from the coordinates $(x, y)$ respectively. Next, we perform a rotation about the temporary origin, and finally, we add $(d_x, d_y)$ back to the rotated point to compensate for the original subtraction. Here are the steps:

1. Subtract $(d_x, d_y)$ to create a new temporary origin:

$$x_1 = (x - d_x)$$
$$y_1 = (y - d_y).$$

2. Rotate $(x_1, y_1)$ about the temporary origin by $\beta$:

$$x_2 = (x - d_x)\cos\beta - (y - d_y)\sin\beta$$
$$y_2 = (x - d_x)\sin\beta + (y - d_y)\cos\beta.$$

3. Add $(d_x, d_y)$ to the rotated point $(x_2, y_2)$ to return to the original origin:

$$x' = x_2 + d_x$$
$$y' = y_2 + d_y$$
$$x' = (x - d_x)\cos\beta - (y - d_y)\sin\beta + d_x$$
$$y' = (x - d_x)\sin\beta + (y - d_y)\cos\beta + d_y.$$

Simplifying, we obtain

$$x' = x\cos\beta - y\sin\beta + d_x(1-\cos\beta) + d_y\sin\beta$$
$$y' = x\sin\beta + y\cos\beta + d_y(1-\cos\beta) - d_x\sin\beta$$

and in matrix form we have

$$\begin{bmatrix} x' \\ y' \\ 1 \end{bmatrix} = \begin{bmatrix} \cos\beta & -\sin\beta & d_x(1-\cos\beta)+d_y\sin\beta \\ \sin\beta & \cos\beta & d_y(1-\cos\beta)-d_x\sin\beta \\ 0 & 0 & 1 \end{bmatrix} \begin{bmatrix} x \\ y \\ 1 \end{bmatrix}. \tag{5.7}$$

For example, if we rotate the point (2, 1), 90° about the point (1, 1), (5.7) becomes

$$\begin{bmatrix} 1 \\ 2 \\ 1 \end{bmatrix} = \begin{bmatrix} 0 & -1 & 2 \\ 1 & 0 & 0 \\ 0 & 0 & 1 \end{bmatrix} \begin{bmatrix} 2 \\ 1 \\ 1 \end{bmatrix},$$

which is confirmed.

The above algebraic approach to derive the rotation transform is relatively easy. However, it is also possible to use matrices to derive composite transforms, such as a reflection relative to an arbitrary line or scaling and rotation relative to an arbitrary point. All of these linear transforms are called *affine* transforms, as parallel lines remain parallel after being transformed. Furthermore, the word 'affine' is used to imply that there is a strong geometric *affinity* between the original and transformed shape. One can not always guarantee that angles and lengths are preserved, as these can change when different scaling factors are used. For completeness, let's derive this transform using matrices.

The homogeneous transform for rotating a point $\beta$ about the origin is given by

$$\mathbf{R}_\beta = \begin{bmatrix} \cos\beta & -\sin\beta & 0 \\ \sin\beta & \cos\beta & 0 \\ 0 & 0 & 1 \end{bmatrix},$$

and a transform for translating a point $(d_x, d_y)$ relative to the origin is given by

$$\mathbf{T}_{d_x,d_y} = \begin{bmatrix} 1 & 0 & d_x \\ 0 & 1 & d_y \\ 0 & 0 & 1 \end{bmatrix}.$$

We can use $\mathbf{R}_\beta$ and $\mathbf{T}_{d_x,d_y}$ to develop a composite transform for rotating a point about an arbitrary point $(d_x, d_y)$ as follows:

$$\begin{bmatrix} x' \\ y' \\ 1 \end{bmatrix} = [\mathbf{T}_{d_x,d_y}][\mathbf{R}_\beta][\mathbf{T}_{-d_x,-d_y}] \begin{bmatrix} x \\ y \\ 1 \end{bmatrix}, \tag{5.8}$$

where

$[\mathbf{T}_{-d_x,-d_y}]$ creates a temporary origin
$[\mathbf{R}_\beta]$ rotates $\beta$ about the temporary origin
$[\mathbf{T}_{d_x,d_y}]$ returns to the original position.

Note that the transform sequence starts on the right next to the original coordinates, working leftwards. Equation (5.8) expands to

$$\begin{bmatrix} x' \\ y' \\ 1 \end{bmatrix} = \begin{bmatrix} 1 & 0 & d_x \\ 0 & 1 & d_y \\ 0 & 0 & 1 \end{bmatrix} \begin{bmatrix} \cos\beta & -\sin\beta & 0 \\ \sin\beta & \cos\beta & 0 \\ 0 & 0 & 1 \end{bmatrix} \begin{bmatrix} 1 & 0 & -d_x \\ 0 & 1 & -d_y \\ 0 & 0 & 1 \end{bmatrix} \begin{bmatrix} x \\ y \\ 1 \end{bmatrix}.$$

Next, we multiply these matrices together to form a single matrix. Let's begin by multiplying the $\mathbf{R}_\beta$ and $\mathbf{T}_{-d_x,-d_y}$ matrices, which produces

$$\begin{bmatrix} x' \\ y' \\ 1 \end{bmatrix} = \begin{bmatrix} 1 & 0 & d_x \\ 0 & 1 & d_y \\ 0 & 0 & 1 \end{bmatrix} \begin{bmatrix} \cos\beta & -\sin\beta & -d_x\cos\beta + d_y\sin\beta \\ \sin\beta & \cos\beta & -d_x\sin\beta - d_y\cos\beta \\ 0 & 0 & 1 \end{bmatrix} \begin{bmatrix} x \\ y \\ 1 \end{bmatrix},$$

and finally we obtain

$$\begin{bmatrix} x' \\ y' \\ 1 \end{bmatrix} = \begin{bmatrix} \cos\beta & -\sin\beta & d_x(1-\cos\beta) + d_y\sin\beta \\ \sin\beta & \cos\beta & d_y(1-\cos\beta) - d_x\sin\beta \\ 0 & 0 & 1 \end{bmatrix} \begin{bmatrix} x \\ y \\ 1 \end{bmatrix},$$

which is the same as the previous transform (5.7).

### 5.7.2 *Rotation and Translation*

There are two ways we can combine the rotate and translate transforms into a single transform. The first way starts by translating a point $P(x, y)$ using $\mathbf{T}_{d_x,d_y}$ to an intermediate point $P''(x + d_x, y + d_y)$ and then rotating this using $\mathbf{R}_\beta$. The problem with this strategy is that the radius of rotation becomes large and subjects the point to a large circular motion. The normal way is to first subject the point to a rotation about the origin and then translate it:

$$P' = [\mathbf{T}_{d_x,d_y}][\mathbf{R}_\beta]P$$

$$\begin{bmatrix} x' \\ y' \\ 1 \end{bmatrix} = \begin{bmatrix} 1 & 0 & d_x \\ 0 & 1 & d_y \\ 0 & 0 & 1 \end{bmatrix} \begin{bmatrix} \cos\beta & -\sin\beta & 0 \\ \sin\beta & \cos\beta & 0 \\ 0 & 0 & 1 \end{bmatrix} \begin{bmatrix} x \\ y \\ 1 \end{bmatrix}$$

$$\begin{bmatrix} x' \\ y' \\ 1 \end{bmatrix} = \begin{bmatrix} \cos\beta & -\sin\beta & d_x \\ \sin\beta & \cos\beta & d_y \\ 0 & 0 & 1 \end{bmatrix} \begin{bmatrix} x \\ y \\ 1 \end{bmatrix}.$$

For example, consider rotating the point $P(1,0)$, $90^\circ$ and then translating it by $(1,0)$. The rotation moves $P$ to $(0,1)$ and the translation moves it to $(1,1)$. This is confirmed by the above transform:

$$\begin{bmatrix} 1 \\ 1 \\ 1 \end{bmatrix} = \begin{bmatrix} 0 & -1 & 1 \\ 1 & 0 & 0 \\ 0 & 0 & 1 \end{bmatrix} \begin{bmatrix} 1 \\ 0 \\ 1 \end{bmatrix}.$$

### 5.7.3 Composite Rotations

It is worth confirming that if we rotate a point $\beta$ about the origin, and follow this by a rotation of $\theta$, this is equivalent to a single rotation of the point through an angle $\theta + \beta$. Let's start with the transforms $\mathbf{R}_\beta$ and $\mathbf{R}_\theta$:

$$\mathbf{R}_\beta = \begin{bmatrix} \cos\beta & -\sin\beta \\ \sin\beta & \cos\beta \end{bmatrix}$$

$$\mathbf{R}_\theta = \begin{bmatrix} \cos\theta & -\sin\theta \\ \sin\theta & \cos\theta \end{bmatrix}.$$

We can represent the double rotation by the product $\mathbf{R}_\theta \mathbf{R}_\beta$:

$$\begin{aligned} \mathbf{R}_\theta \mathbf{R}_\beta &= \begin{bmatrix} \cos\theta & -\sin\theta \\ \sin\theta & \cos\theta \end{bmatrix} \begin{bmatrix} \cos\beta & -\sin\beta \\ \sin\beta & \cos\beta \end{bmatrix} \\ &= \begin{bmatrix} \cos\theta\cos\beta - \sin\theta\sin\beta & -\cos\theta\sin\beta - \sin\theta\cos\beta \\ \sin\theta\cos\beta + \cos\theta\sin\beta & -\sin\theta\sin\beta + \cos\theta\cos\beta \end{bmatrix} \\ &= \begin{bmatrix} \cos(\theta+\beta) & -\sin(\theta+\beta) \\ \sin(\theta+\beta) & \cos(\theta+\beta) \end{bmatrix} \end{aligned}$$

which confirms that the composite rotation is equivalent to a single rotation of $\theta + \beta$.

## 5.8 Change of Axes

Points in one coordinate system often have to be referenced in another one. For example, to view a 3D scene from an arbitrary position, a virtual camera is positioned in the world space using a series of transforms. An object's coordinates, which are relative to the world frame of reference, are computed relative to the camera's axial system, and then used to develop a perspective projection. Let's examine how one changes axial systems in two dimensions.

Figure 5.9 shows a point $P(x, y)$ relative to the $xy$ axes, but we require to know the coordinates relative to the $x'y'$ axes. To do this, we need to know the relationship between the two coordinate systems, and ideally we want to apply a technique

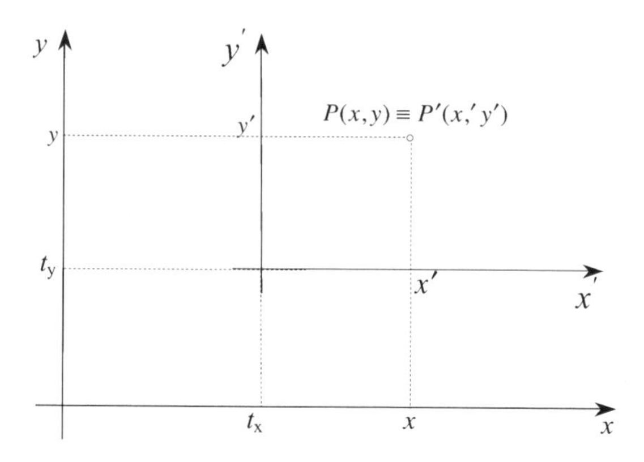


**Fig. 5.9** The $x'y'$ axial system is translated $(d_x, d_y)$

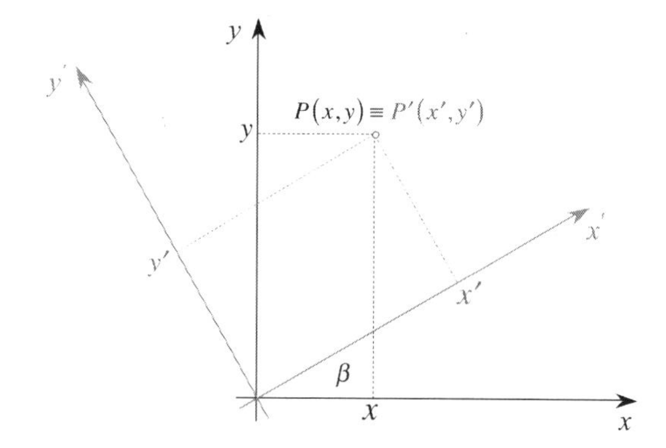


**Fig. 5.10** The $x'y'$ axial system is rotated $\beta$

that works in 2D and 3D. If the second coordinate system is a simple translation $(d_x, d_y)$ relative to the reference system, as shown in Fig. 5.9, the point $P(x, y)$ has coordinates relative to the translated system $(x - d_x, y - d_y)$:

$$\begin{bmatrix} x' \\ y' \\ 1 \end{bmatrix} = \begin{bmatrix} 1 & 0 & -d_x \\ 0 & 1 & -d_y \\ 0 & 0 & 1 \end{bmatrix} \begin{bmatrix} x \\ y \\ 1 \end{bmatrix}.$$

If the $x'y'$ axes are rotated $\beta$ relative to the $xy$ axes, as shown in Fig. 5.10, a point $P(x, y)$ relative to the $xy$ axes becomes $P'(x', y')$ relative to the rotated axes is given by

$$\begin{bmatrix} x' \\ y' \\ 1 \end{bmatrix} = \begin{bmatrix} \cos(-\beta) & -\sin(-\beta) & 0 \\ \sin(-\beta) & \cos(-\beta) & 0 \\ 0 & 0 & 1 \end{bmatrix} \begin{bmatrix} x \\ y \\ 1 \end{bmatrix},$$

which simplifies to

$$\begin{bmatrix} x' \\ y' \\ 1 \end{bmatrix} = \begin{bmatrix} \cos\beta & \sin\beta & 0 \\ -\sin\beta & \cos\beta & 0 \\ 0 & 0 & 1 \end{bmatrix} \begin{bmatrix} x \\ y \\ 1 \end{bmatrix}.$$

When a coordinate system is rotated and translated relative to the reference system, a point $P(x, y)$ becomes $P'(x', y')$ relative to the new axes given by

$$\begin{bmatrix} x' \\ y' \\ 1 \end{bmatrix} = \begin{bmatrix} \cos\beta & \sin\beta & 0 \\ -\sin\beta & \cos\beta & 0 \\ 0 & 0 & 1 \end{bmatrix} \begin{bmatrix} 1 & 0 & -d_x \\ 0 & 1 & -d_y \\ 0 & 0 & 1 \end{bmatrix} \begin{bmatrix} x \\ y \\ 1 \end{bmatrix},$$

which simplifies to

$$\begin{bmatrix} x' \\ y' \\ 1 \end{bmatrix} = \begin{bmatrix} \cos\beta & \sin\beta & -d_x\cos\beta - d_y\sin\beta \\ -\sin\beta & \cos\beta & d_x\sin\beta - d_y\cos\beta \\ 0 & 0 & 1 \end{bmatrix} \begin{bmatrix} x \\ y \\ 1 \end{bmatrix}.$$

## 5.9 Eigenvectors and Eigenvalues

Figure 5.11 shows the result of applying the following transform to a unit square. There is a pronounced stretching in the first and third quadrants, and reduced stretching in the second and fourth quadrants.

$$\begin{bmatrix} x' \\ y' \end{bmatrix} = \begin{bmatrix} 4 & 1 \\ 1 & 4 \end{bmatrix} \begin{bmatrix} x \\ y \end{bmatrix}.$$

It should be clear from Fig. 5.11 that any point $(k, k)$ is transformed to another point $(5k, 5k)$, and that its mirror point $(-k, -k)$ is transformed to $(-5k, -5k)$. Similarly, any point $(-k, k)$ is transformed to another point $(-3k, 3k)$, and its mirror point $(k, -k)$ is transformed to $(3k, -3k)$. Thus the transform shows a particular bias towards points lying on vectors $[k \quad k]^T$ and $[-k \quad k]^T$, where $k \neq 0$. These vectors are called *eigenvectors* and the scaling factor is its *eigenvalue*. Figure 5.12 shows a scenario where a transform $t$ moves point $R$ to $S$, whilst the same transform moves $P$—which lies on one of $t$'s eigenvectors, to $Q$—which also lies on the same eigenvector.

We can define an eigenvector and its eigenvalue as follows. Given a square matrix $\mathbf{A}$, a non-zero vector $\mathbf{v}$ is an eigenvector, and $\lambda$ is the corresponding eigenvalue if

$$\mathbf{Av} = \lambda\mathbf{v},$$

where $\lambda$ is a scalar.

The German word *eigen* means *characteristic*, *own*, *latent* or *special*, and eigenvector means a special vector associated with a transform. The equation that determines the existence of any eigenvectors is called the *characteristic equation* of a square matrix, and is given by

$$|\mathbf{A} - \lambda\mathbf{I}| = 0. \tag{5.9}$$

Let's derive the characteristic equation (5.9).

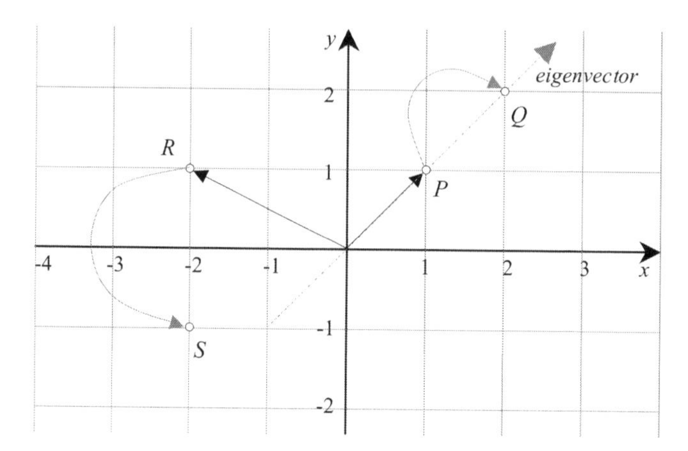


**Fig. 5.11** Transforming points on four unit squares

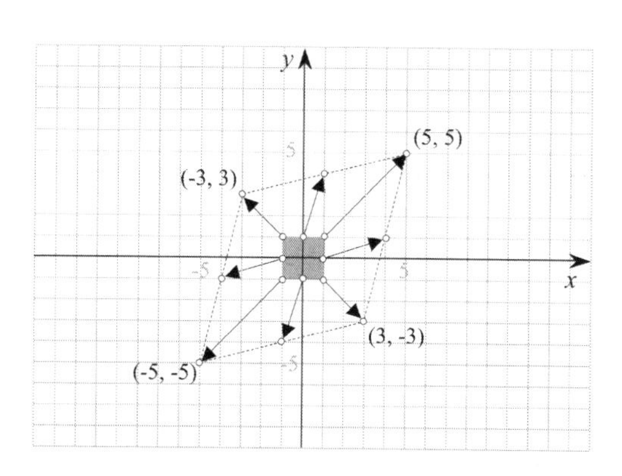


**Fig. 5.12** How a transform reacts to different points

Consider the 2D transform $t$ that maps the point $(x, y)$ to another point $(ax + by, cx + dy)$:

$$t(x, y) \mapsto (ax + by, cx + dy).$$

This is expressed in matrix form as

$$t : \mathbf{v} \mapsto \mathbf{A}\mathbf{v},$$

or

$$\begin{bmatrix} x' \\ y' \end{bmatrix} = \begin{bmatrix} a & b \\ c & d \end{bmatrix} \begin{bmatrix} x \\ y \end{bmatrix},$$

where

$$\mathbf{A} = \begin{bmatrix} a & b \\ c & d \end{bmatrix}, \qquad \mathbf{v} = \begin{bmatrix} x \\ y \end{bmatrix}.$$

Therefore, if $\mathbf{v}$ is an eigenvector of $t$, and $\lambda$ its associated eigenvalue, then

$$\mathbf{A}\mathbf{v} = \lambda\mathbf{v}$$

$$\begin{bmatrix} a & b \\ c & d \end{bmatrix}\begin{bmatrix} x \\ y \end{bmatrix} = \lambda\begin{bmatrix} x \\ y \end{bmatrix}$$

or in equation terms:

$$ax + by = \lambda x$$
$$cx + dy = \lambda y.$$

Rearranging, we have

$$(a-\lambda)x + by = 0$$
$$cx + (d-\lambda)y = 0$$

or back in matrix form:

$$\begin{bmatrix} a-\lambda & b \\ c & d-\lambda \end{bmatrix}\begin{bmatrix} x \\ y \end{bmatrix} = \begin{bmatrix} 0 \\ 0 \end{bmatrix}.$$

For a non-zero $[x \quad y]^{\mathrm{T}}$ to exist, we must have

$$\begin{vmatrix} a-\lambda & b \\ c & d-\lambda \end{vmatrix} = 0,$$

which is called the *characteristic equation*. Let's use this on the transform

$$\begin{bmatrix} 4 & 1 \\ 1 & 4 \end{bmatrix}\begin{bmatrix} x \\ y \end{bmatrix} = \begin{bmatrix} x' \\ y' \end{bmatrix}.$$

Then

$$\begin{vmatrix} 4-\lambda & 1 \\ 1 & 4-\lambda \end{vmatrix} = 0$$
$$(4-\lambda)^2 - 1 = 0$$
$$\lambda^2 - 8\lambda + 16 - 1 = 0$$
$$\lambda^2 - 8\lambda + 15 = 0$$
$$(\lambda-5)(\lambda-3) = 0.$$

Thus $\lambda = 5$ and $\lambda = 3$, are the two eigenvalues we observed in Fig. 5.11. Next, we substitute the two values of $\lambda$ in

$$\begin{bmatrix} 4-\lambda & 1 \\ 1 & 4-\lambda \end{bmatrix}\begin{bmatrix} x \\ y \end{bmatrix} = \begin{bmatrix} 0 \\ 0 \end{bmatrix},$$

to extract the eigenvectors. Let's start with $\lambda = 5$:

$$\begin{bmatrix} -1 & 1 \\ 1 & -1 \end{bmatrix} \begin{bmatrix} x \\ y \end{bmatrix} = \begin{bmatrix} 0 \\ 0 \end{bmatrix},$$

which represents the equation $y = x$ or the vector $[k \quad k]^{\mathrm{T}}$. Next, we substitute $\lambda = 3$:

$$\begin{bmatrix} 1 & 1 \\ 1 & 1 \end{bmatrix} \begin{bmatrix} x \\ y \end{bmatrix} = \begin{bmatrix} 0 \\ 0 \end{bmatrix},$$

which represents the equation $y = -x$ or the vector $[-k \quad k]^{\mathrm{T}}$. Thus we have discovered that the transform possesses two eigenvectors $[k \quad k]^{\mathrm{T}}$ and $[-k \quad k]^{\mathrm{T}}$ and their respective eigenvalues $\lambda = 5$ and $\lambda = 3$, as predicted.

The characteristic equation may have real or complex solutions, and if they are complex, there are no real eigenvectors. For example, we would not expect the following rotation transform to have any real eigenvectors, as this would imply that it shows a rotational preference to certain points. Let's explore this transform to see how the characteristic equation behaves.

$$\mathbf{R} = \begin{bmatrix} \cos\beta & -\sin\beta \\ \sin\beta & \cos\beta \end{bmatrix}.$$

The characteristic equation is

$$\begin{vmatrix} \cos\beta - \lambda & -\sin\beta \\ \sin\beta & \cos\beta - \lambda \end{vmatrix} = 0,$$

where $\beta$ is the angle of rotation. Therefore,

$$\begin{aligned} (\cos\beta - \lambda)^2 + \sin^2\beta &= 0 \\ \lambda^2 - 2\lambda\cos\beta + \cos^2\beta + \sin^2\beta &= 0 \\ \lambda^2 - 2\lambda\cos\beta + 1 &= 0. \end{aligned}$$

This quadratic in $\lambda$ is solved using

$$\lambda = \frac{-b \pm \sqrt{b^2 - 4ac}}{2a},$$

where $a = 1$, $b = -2\cos\beta$, $c = 1$:

$$\begin{aligned} \lambda &= \frac{2\cos\beta \pm \sqrt{4\cos^2\beta - 4}}{2} \\ &= \cos\beta \pm \sqrt{\cos^2\beta - 1} \\ &= \cos\beta \pm \sqrt{-\sin^2\beta} \end{aligned}$$

$$\lambda_1 = \cos\beta + i \sin\beta$$
$$\lambda_2 = \cos\beta - i \sin\beta$$

which are complex numbers.

The corresponding complex eigenvectors are

$$\mathbf{v}_1 = \begin{bmatrix} 1 \\ i \end{bmatrix}$$
$$\mathbf{v}_2 = \begin{bmatrix} 1 \\ -i \end{bmatrix}.$$

The major problem with the above technique is that it requires careful analysis to untangle the eigenvector, and ideally, we require a deterministic algorithm to reveal the result. In Chap. 6 we will discover such a technique.

## 5.10 Worked Examples

*Example 1* State the matrix transform to translate $-10$ units in the $x$ direction and 5 units in the $y$ direction.

$$\begin{bmatrix} 1 & 0 & -10 \\ 0 & 1 & 5 \\ 0 & 0 & 1 \end{bmatrix}.$$

*Example 2* State the matrix transform to scale 2 in the $x$ direction and 4 in the $y$ direction.

$$\begin{bmatrix} 2 & 0 \\ 0 & 4 \end{bmatrix}.$$

*Example 3* Derive the matrix transform to scale 2 relative to the point $(1, 1)$, and show that the point $(1, 1)$ is not moved.

$$\begin{aligned} \mathbf{S} &= \mathbf{T}_{(1,1)}\mathbf{S}_{(2)}\mathbf{T}_{(-1,-1)} \\ &= \begin{bmatrix} 1 & 0 & 1 \\ 0 & 1 & 1 \\ 0 & 0 & 1 \end{bmatrix} \begin{bmatrix} 2 & 0 & 0 \\ 0 & 2 & 0 \\ 0 & 0 & 1 \end{bmatrix} \begin{bmatrix} 1 & 0 & -1 \\ 0 & 1 & -1 \\ 0 & 0 & 1 \end{bmatrix} \\ &= \begin{bmatrix} 2 & 0 & 1 \\ 0 & 2 & 1 \\ 0 & 0 & 1 \end{bmatrix} \begin{bmatrix} 1 & 0 & -1 \\ 0 & 1 & -1 \\ 0 & 0 & 1 \end{bmatrix} \\ &= \begin{bmatrix} 2 & 0 & -1 \\ 0 & 2 & -1 \\ 0 & 0 & 1 \end{bmatrix} \end{aligned}$$

$$\begin{bmatrix} 1 \\ 1 \\ 1 \end{bmatrix} = \begin{bmatrix} 2 & 0 & -1 \\ 0 & 2 & -1 \\ 0 & 0 & 1 \end{bmatrix} \begin{bmatrix} 1 \\ 1 \\ 1 \end{bmatrix}.$$

*Example 4* Use the reflection transform to reflect the point (0, 1) about the vector $[-1 \quad 1]^T$.

$$\begin{bmatrix} x' \\ y' \end{bmatrix} = \begin{bmatrix} u_x^2 - u_y^2 & 2u_x u_y \\ 2u_x u_y & u_y^2 - u_x^2 \end{bmatrix} \begin{bmatrix} x \\ y \end{bmatrix},$$

where $u_x = -\frac{\sqrt{2}}{2}$, $u_y = \frac{\sqrt{2}}{2}$

$$\begin{bmatrix} -1 \\ 0 \end{bmatrix} = \begin{bmatrix} 0 & -1 \\ -1 & 0 \end{bmatrix} \begin{bmatrix} 0 \\ 1 \end{bmatrix},$$

which is correct.

*Example 5* Rotate the point $(1, 0)$ $30°$ about the origin.

$$\begin{bmatrix} x' \\ y' \end{bmatrix} = \begin{bmatrix} \cos 30° & -\sin 30° \\ \sin 30° & \cos 30° \end{bmatrix} \begin{bmatrix} 1 \\ 0 \end{bmatrix}$$

$$\begin{bmatrix} \frac{\sqrt{3}}{2} \\ \frac{1}{2} \end{bmatrix} = \begin{bmatrix} \frac{\sqrt{3}}{2} & -\frac{1}{2} \\ \frac{1}{2} & \frac{\sqrt{3}}{2} \end{bmatrix} \begin{bmatrix} 1 \\ 0 \end{bmatrix}.$$

The rotated point is $(\frac{\sqrt{3}}{2}, \frac{1}{2})$.

*Example 6* What happens to the unit square when it is transformed by a negative, unit scaling matrix?

Let the unit-square coordinates be in an counter-clockwise sequence: $(0, 0)$, $(1, 0)$, $(1, 1)$, $(0, 1)$.

Let the negative, unit scaling matrix be

$$\begin{bmatrix} -1 & 0 \\ 0 & -1 \end{bmatrix}.$$

Therefore,

$$\begin{bmatrix} 0 \\ 0 \end{bmatrix} = \begin{bmatrix} -1 & 0 \\ 0 & -1 \end{bmatrix} \begin{bmatrix} 0 \\ 0 \end{bmatrix}$$

$$\begin{bmatrix} -1 \\ 0 \end{bmatrix} = \begin{bmatrix} -1 & 0 \\ 0 & -1 \end{bmatrix} \begin{bmatrix} 1 \\ 0 \end{bmatrix}$$

$$\begin{bmatrix} -1 \\ -1 \end{bmatrix} = \begin{bmatrix} -1 & 0 \\ 0 & -1 \end{bmatrix} \begin{bmatrix} 1 \\ 1 \end{bmatrix}$$

$$\begin{bmatrix} 0 \\ -1 \end{bmatrix} = \begin{bmatrix} -1 & 0 \\ 0 & -1 \end{bmatrix} \begin{bmatrix} 0 \\ 1 \end{bmatrix}.$$

By inspection, the unit square is rotated 180°. Furthermore, the rotation matrix is

$$\begin{bmatrix} \cos\theta & -\sin\theta \\ \sin\theta & \cos\theta \end{bmatrix},$$

and when $\theta = 180°$, the matrix becomes

$$\begin{bmatrix} -1 & 0 \\ 0 & -1 \end{bmatrix}.$$

*Example 7* Rotate the unit square in Example 6 180° about the point (1, 1).
The matrix transform for rotating about an arbitrary point is

$$\begin{bmatrix} x' \\ y' \\ 1 \end{bmatrix} = \begin{bmatrix} \cos\beta & -\sin\beta & d_x(1-\cos\beta) + d_y \sin\beta \\ \sin\beta & \cos\beta & d_y(1-\cos\beta) - d_x \sin\beta \\ 0 & 0 & 1 \end{bmatrix} \begin{bmatrix} x \\ y \\ 1 \end{bmatrix},$$

where $(d_x, d_y)$ is the point of rotation. With $\beta = 180°$:

$$\begin{bmatrix} x' \\ y' \\ 1 \end{bmatrix} = \begin{bmatrix} -1 & 0 & 2 \\ 0 & -1 & 2 \\ 0 & 0 & 1 \end{bmatrix} \begin{bmatrix} x \\ y \\ 1 \end{bmatrix}.$$

Substituting the unit-square coordinates, we have:

$$\begin{bmatrix} 2 \\ 2 \\ 1 \end{bmatrix} = \begin{bmatrix} -1 & 0 & 2 \\ 0 & -1 & 2 \\ 0 & 0 & 1 \end{bmatrix} \begin{bmatrix} 0 \\ 0 \\ 1 \end{bmatrix}$$

$$\begin{bmatrix} 1 \\ 2 \\ 1 \end{bmatrix} = \begin{bmatrix} -1 & 0 & 2 \\ 0 & -1 & 2 \\ 0 & 0 & 1 \end{bmatrix} \begin{bmatrix} 1 \\ 0 \\ 1 \end{bmatrix}$$

$$\begin{bmatrix} 1 \\ 1 \\ 1 \end{bmatrix} = \begin{bmatrix} -1 & 0 & 2 \\ 0 & -1 & 2 \\ 0 & 0 & 1 \end{bmatrix} \begin{bmatrix} 1 \\ 1 \\ 1 \end{bmatrix}$$

$$\begin{bmatrix} 2 \\ 1 \\ 1 \end{bmatrix} = \begin{bmatrix} -1 & 0 & 2 \\ 0 & -1 & 2 \\ 0 & 0 & 1 \end{bmatrix} \begin{bmatrix} 0 \\ 1 \\ 1 \end{bmatrix}.$$

## 5.11 Summary

In this chapter we have studied various geometric transforms and used algebraic expressions to derive their matrix equivalent. Hopefully, it is becoming clear that matrix notation offers a powerful notation for combining these transforms, without resorting to algebra. Eventually, you should be able to recognise a transform simply from the pattern of its elements.

The determinant of a matrix controls the area change that a shape undergoes after being transformed. Often this is unity, which results in no change of area. Naturally, scaling must introduce a change in area. Reflection, on the other hand, reverses a shape's vertex sequence.

Lastly, this chapter showed how it is possible to discover whether a matrix possesses any preferred directions; these are called its eigenvectors and associated eigenvalues. We will develop these ideas ion the following chapter.

# Chapter 6
# 3D Transforms

## 6.1 Introduction

In this chapter we generalise the techniques of 2D transforms into a 3D context, where scaling, shearing, translation and reflection are very similar to their 2D counterparts, but rotation transforms are complicated by the number of combinations that arise when rotations about the Cartesian axes are combined together. Because these composite rotations are flawed, we develop a rotation transform about an arbitrary axis using vectors and another using matrices.

## 6.2 Scaling

3D scaling is represented algebraically as

$$\begin{aligned} x' &= s_x x \\ y' &= s_y y \\ z' &= s_z z \end{aligned}$$

or in matrix form:

$$\begin{bmatrix} x' \\ y' \\ z' \end{bmatrix} = \begin{bmatrix} s_x & 0 & 0 \\ 0 & s_y & 0 \\ 0 & 0 & s_z \end{bmatrix} \begin{bmatrix} x \\ y \\ z \end{bmatrix}.$$

The scaling is relative to the origin, and the determinant $s_x s_y s_z$ represents the change in volume an object undergoes after the transform. The inverse transform $\mathbf{S}^{-1}$ is given by

J. Vince, *Matrix Transforms for Computer Games and Animation*,
DOI 10.1007/978-1-4471-4321-5_6, 

$$\mathbf{S}^{-1} = \begin{bmatrix} \frac{1}{s_x} & 0 & 0 \\ 0 & \frac{1}{s_y} & 0 \\ 0 & 0 & \frac{1}{s_z} \end{bmatrix},$$

because

$$\mathbf{SS}^{-1} = \begin{bmatrix} s_x & 0 & 0 \\ 0 & s_y & 0 \\ 0 & 0 & s_z \end{bmatrix} \begin{bmatrix} \frac{1}{s_x} & 0 & 0 \\ 0 & \frac{1}{s_y} & 0 \\ 0 & 0 & \frac{1}{s_z} \end{bmatrix} = \begin{bmatrix} 1 & 0 & 0 \\ 0 & 1 & 0 \\ 0 & 0 & 1 \end{bmatrix}.$$

We can also arrange for it to be relative to an arbitrary point $(p_x, p_y, p_z)$ by using the same scenario used in 2D:

$$\begin{aligned} x' &= s_x(x - p_x) + p_x \\ y' &= s_y(y - p_y) + p_y \\ z' &= s_z(z - p_z) + p_z \end{aligned}$$

which as a homogeneous matrix is

$$\begin{bmatrix} x' \\ y' \\ z' \\ 1 \end{bmatrix} = \begin{bmatrix} s_x & 0 & 0 & p_x(1 - s_x) \\ 0 & s_y & 0 & p_y(1 - s_y) \\ 0 & 0 & s_z & p_z(1 - s_z) \\ 0 & 0 & 0 & 1 \end{bmatrix} \begin{bmatrix} x \\ y \\ z \\ 1 \end{bmatrix}.$$

Observe that the determinant is still $s_x s_y s_z$.

## 6.3 Translation

Knowing the matrix for 2D translation, it is a trivial step to write directly the 3D equivalent:

$$\begin{bmatrix} x' \\ y' \\ z' \\ 1 \end{bmatrix} = \begin{bmatrix} 1 & 0 & 0 & d_x \\ 0 & 1 & 0 & d_y \\ 0 & 0 & 1 & d_z \\ 0 & 0 & 0 & 1 \end{bmatrix} \begin{bmatrix} x \\ y \\ z \\ 1 \end{bmatrix}.$$

The determinant remains equal to 1. The inverse is:

$$\begin{bmatrix} 1 & 0 & 0 & -d_x \\ 0 & 1 & 0 & -d_y \\ 0 & 0 & 1 & -d_z \\ 0 & 0 & 0 & 1 \end{bmatrix}.$$

## 6.4 Shearing

In 3D there are three axes to shear an object, which means that each shear along an axis is parallel with one of the two planes sharing the axis. For example, the $x$ axis shear is determined either by the $y$ coordinate of a vertex, or its $z$ coordinate, which results in 6 transforms:

$$\mathbf{S}_{x_y} = \begin{bmatrix} 1 & \tan\alpha & 0 \\ 0 & 1 & 0 \\ 0 & 0 & 1 \end{bmatrix}, \qquad \mathbf{S}_{x_z} = \begin{bmatrix} 1 & 0 & \tan\alpha \\ 0 & 1 & 0 \\ 0 & 0 & 1 \end{bmatrix}$$

$$\mathbf{S}_{y_x} = \begin{bmatrix} 1 & 0 & 0 \\ \tan\alpha & 1 & 0 \\ 0 & 0 & 1 \end{bmatrix}, \qquad \mathbf{S}_{y_z} = \begin{bmatrix} 1 & 0 & 0 \\ 0 & 1 & \tan\alpha \\ 0 & 0 & 1 \end{bmatrix}$$

$$\mathbf{S}_{z_x} = \begin{bmatrix} 1 & 0 & 0 \\ 0 & 1 & 0 \\ \tan\alpha & 0 & 1 \end{bmatrix}, \qquad \mathbf{S}_{z_y} = \begin{bmatrix} 1 & 0 & 0 \\ 0 & 1 & 0 \\ 0 & \tan\alpha & 1 \end{bmatrix}.$$

## 6.5 Reflection in a Plane Intersecting the Origin

Reflecting points in a plane sounds slightly daunting, but it is not; for having derived a 2D solution, the 3D solution is identical. Figure 6.1 shows a plane intersecting the origin containing $\mathbf{u}$, and a unit normal vector $\mathbf{n} = a\mathbf{i} + b\mathbf{j} + c\mathbf{k}$. We now have exactly the same geometry used to solve the 2D case, apart from the vectors being 3D. In the 2D case the transform was given by

$$\mathbf{p}' = (\mathbf{I} - 2\mathbf{n}\mathbf{n}^{\mathrm{T}})\mathbf{p},$$

where $\mathbf{R} = \mathbf{I} - 2\mathbf{n}\mathbf{n}^{\mathrm{T}}$ is the transform, which in a 3D scenario becomes:

$$\begin{aligned} \mathbf{R} &= \begin{bmatrix} 1 & 0 & 0 \\ 0 & 1 & 0 \\ 0 & 0 & 1 \end{bmatrix} - 2\begin{bmatrix} a \\ b \\ c \end{bmatrix} [a \quad b \quad c] \\ &= \begin{bmatrix} 1 & 0 & 0 \\ 0 & 1 & 0 \\ 0 & 0 & 1 \end{bmatrix} - 2\begin{bmatrix} a^2 & ab & ac \\ ab & b^2 & bc \\ ac & bc & c^2 \end{bmatrix} \\ &= \begin{bmatrix} 1-2a^2 & -2ab & -2ac \\ -2ab & 1-2b^2 & -2bc \\ -2ac & -2bc & 1-2c^2 \end{bmatrix}. \end{aligned}$$

Let's compute the determinant of $\mathbf{R}$ using Sarrus' rule to show that it equals $-1$:

$$|\mathbf{R}| = \begin{vmatrix} 1-2a^2 & -2ab & -2ac \\ -2ab & 1-2b^2 & -2bc \\ -2ac & -2bc & 1-2c^2 \end{vmatrix}$$

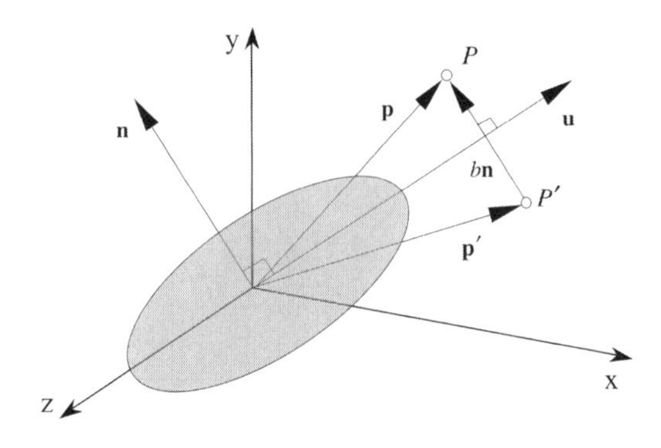


**Fig. 6.1** Reflecting a point in a plane

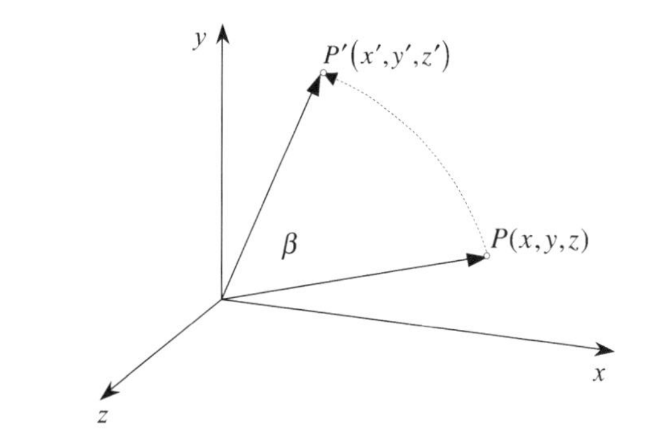


**Fig. 6.2** Rotating the point $P$ about the $z$ axis

$$
\begin{aligned}
&= (1-2a^2)(1-2b^2)(1-2c^2) - 8a^2b^2c^2 - 8a^2b^2c^2 \\
&\quad - 4a^2c^2(1-2b^2) - 4a^2b^2(1-2c^2) - 4b^2c^2(1-2a^2) \\
&= (1-2a^2)(1-2b^2)(1-2c^2) - 8a^2b^2c^2 - 8a^2b^2c^2 \\
&\quad - 4a^2c^2 + 8a^2b^2c^2 - 4a^2b^2 + 8a^2b^2c^2 - 4b^2c^2 + 8a^2b^2c^2 \\
&= (1-2a^2-2b^2+4a^2b^2)(1-2c^2) - 4a^2c^2 - 4a^2b^2 - 4b^2c^2 + 8a^2b^2c^2 \\
&= 1-2a^2-2b^2-2c^2 \\
&= 1-2(a^2+b^2+c^2) \\
&= -1.
\end{aligned}
$$

Now let's test this transform with an example. Figure 6.2 shows a scenario where a plane intersects the origin with a normal vector given by $\mathbf{n} = \frac{\sqrt{2}}{2}\mathbf{i} + \frac{\sqrt{2}}{2}\mathbf{j}$. The point to be reflected is $P = (1, 1, 0)$ which makes the reflected point $P' = (-1, -1, 0)$. Therefore, the transform becomes:

$$
a = \frac{\sqrt{2}}{2}, \qquad b = \frac{\sqrt{2}}{2}, \qquad c = 0
$$

$$\mathbf{R} = \begin{bmatrix} 0 & -1 & 0 \\ -1 & 0 & 0 \\ 0 & 0 & 1 \end{bmatrix}$$

$$\begin{bmatrix} -1 \\ -1 \\ 0 \end{bmatrix} = \begin{bmatrix} 0 & -1 & 0 \\ -1 & 0 & 0 \\ 0 & 0 & 1 \end{bmatrix} \begin{bmatrix} 1 \\ 1 \\ 0 \end{bmatrix}.$$

## 6.6 Rotation

Although we talk about rotating points about another point in space, we require more precise information to describe this mathematically. We could, for example, associate a plane with the point of rotation and confine the rotated point to this plane, but it's much easier to visualise an axis perpendicular to this plane, about which the rotation occurs. Unfortunately, the matrix algebra for such an operation starts to become fussy, and ultimately we have seek the help of quaternions. So let us begin this investigation by rotating a point about the three fixed Cartesian axes. Such rotations are called *Euler rotations* after the Swiss mathematician Leonhard Euler.

Recall that the transform for rotating a point about the origin in the plane is given by

$$\mathbf{R}_\beta = \begin{bmatrix} \cos\beta & -\sin\beta \\ \sin\beta & \cos\beta \end{bmatrix}.$$

This can be generalised into a 3D rotation $\mathbf{R}_{\beta,z}$ about the $z$ axis by adding a $z$ coordinate as follows

$$\mathbf{R}_{\beta,z} = \begin{bmatrix} \cos\beta & -\sin\beta & 0 \\ \sin\beta & \cos\beta & 0 \\ 0 & 0 & 1 \end{bmatrix},$$

which is illustrated in Fig. 6.2.

To rotate a point about the $x$ axis, the $x$ coordinate remains constant whilst the $y$ and $z$ coordinates are changed according to the 2D rotation transform. This is expressed algebraically as

$$\begin{aligned} x' &= x \\ y' &= y\cos\beta - z\sin\beta \\ z' &= y\sin\beta + z\cos\beta \end{aligned}$$

or in matrix form as $\mathbf{R}_{\beta,x}$

$$\mathbf{R}_{\beta,x} = \begin{bmatrix} 1 & 0 & 0 \\ 0 & \cos\beta & -\sin\beta \\ 0 & \sin\beta & \cos\beta \end{bmatrix}.$$

To rotate about the $y$ axis, the $y$ coordinate remains constant whilst the $x$ and $z$ coordinates are changed. This is expressed algebraically as

$$x' = z \sin\beta + x \cos\beta$$
$$y' = y$$
$$z' = z \cos\beta - x \sin\beta$$

or in matrix form as $\mathbf{R}_{\beta,y}$

$$\mathbf{R}_{\beta,y} = \begin{bmatrix} \cos\beta & 0 & \sin\beta \\ 0 & 1 & 0 \\ -\sin\beta & 0 & \cos\beta \end{bmatrix}.$$

Note that the matrix terms don't appear to share the symmetry enjoyed by the previous two matrices. Nothing really has gone wrong, it's just the way the axes are paired together to rotate the coordinates. Now let's consider similar rotations about off-set axes parallel to the Cartesian axes.

### *6.6.1 Rotation About an Off-Set Axis*

To begin, let's develop a transform to rotate a point about a fixed axis parallel with the $z$ axis, as shown in Fig. 6.3. The scenario is very reminiscent of the 2D case for rotating a point about an arbitrary point, and the general transform is given by

$$\begin{bmatrix} x' \\ y' \\ z' \\ 1 \end{bmatrix} = [\mathbf{T}_{d_x,d_y,0}][\mathbf{R}_{\beta,z}][\mathbf{T}_{-d_x,-d_y,0}] \begin{bmatrix} x \\ y \\ z \\ 1 \end{bmatrix}$$

where

| | |
|---|---|
| $[\mathbf{T}_{-d_x,-d_y,0}]$ | creates a temporary origin |
| $[\mathbf{R}_{\beta,z}]$ | rotates $\beta$ about the temporary $z$ axis |
| $[\mathbf{T}_{d_x,d_y,0}]$ | returns to the original position. |

and the matrix transform is

$$[\mathbf{T}_{d_x,d_y,0}][\mathbf{R}_{\beta,z}][\mathbf{T}_{-d_x,-d_y,0}] = \begin{bmatrix} \cos\beta & -\sin\beta & 0 & d_x(1-\cos\beta)+d_y\sin\beta \\ \sin\beta & \cos\beta & 0 & d_y(1-\cos\beta)-d_x\sin\beta \\ 0 & 0 & 1 & 0 \\ 0 & 0 & 0 & 1 \end{bmatrix}.$$

Hopefully, you can see the similarity between rotating in 3D and 2D—the $x$ and $y$ coordinates are updated while the $z$ coordinate is held constant. We can now state

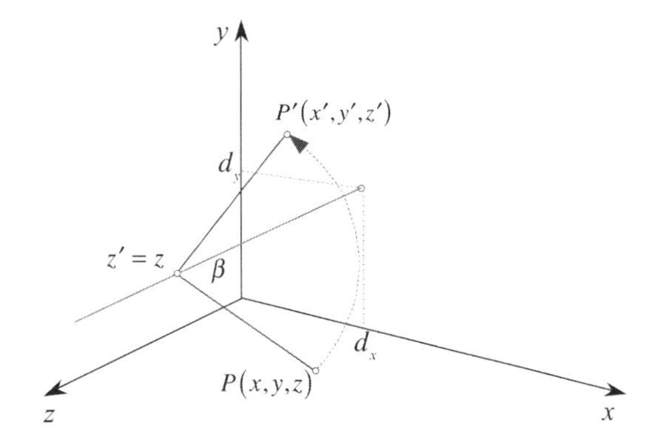


**Fig. 6.3** Rotating a point about an axis parallel with the $z$ axis

the other two matrices for rotating about an off-set axis parallel with the $x$ axis and parallel with the $y$ axis:

$$[\mathbf{T}_{0,d_y,d_z}][\mathbf{R}_{\beta,x}][\mathbf{T}_{0,-d_y,-d_z}] = \begin{bmatrix} 1 & 0 & 0 & 0 \\ 0 & \cos\beta & -\sin\beta & d_y(1-\cos\beta)+d_z\sin\beta \\ 0 & \sin\beta & \cos\beta & d_z(1-\cos\beta)-d_y\sin\beta \\ 0 & 0 & 0 & 1 \end{bmatrix}$$

$$[\mathbf{T}_{d_x,0,d_z}][\mathbf{R}_{\beta,y}][\mathbf{T}_{-d_x,0,-d_z}] = \begin{bmatrix} \cos\beta & 0 & \sin\beta & d_x(1-\cos\beta)-d_z\sin\beta \\ 0 & 1 & 0 & 0 \\ -\sin\beta & 0 & \cos\beta & d_z(1-\cos\beta)+d_x\sin\beta \\ 0 & 0 & 0 & 1 \end{bmatrix}.$$

## *6.6.2 Composite Rotations*

So far we have only considered single rotations about a Cartesian axis or a parallel off-set axis, but there is nothing to stop us constructing a sequence of rotations to create a composite rotation. For example, we could begin by rotating a point $\alpha$ about the $x$ axis followed by a rotation $\beta$ about the $y$ axis, which in turn could be followed by a rotation $\gamma$ about the $z$ axis. As mentioned above, these rotations are called Euler rotations.

One of the problems with Euler rotations is visualising exactly what is happening at each step, and predicting the orientation of an object after a composite rotation. To simplify the problem we will employ a unit cube whose vertices are numbered 0 to 7 as shown in Fig. 6.4. We will also employ the following binary coded expression that uses the Cartesian coordinates of the vertex in the vertex number:

$$vertex = 4x + 2y + z.$$

For example, vertex 0 has coordinates $(0,0,0)$, and vertex 7 has coordinates $(1,1,1)$. All the codes are shown in Table 6.1.

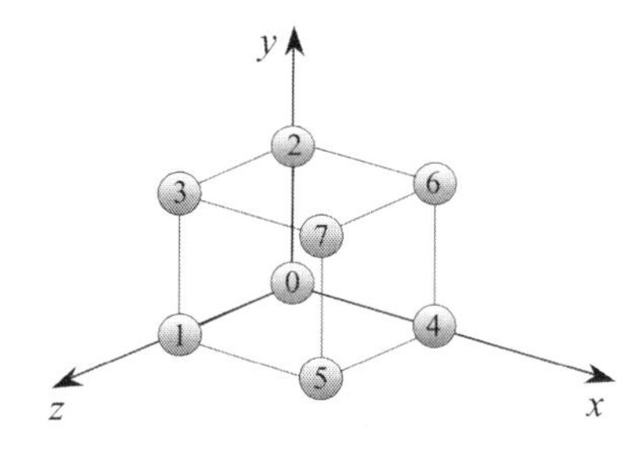


**Fig. 6.4** A unit cube with vertices coded as shown in Table 6.1

**Table 6.1** Vertex coordinates for the cube in Fig. 6.4

| *vertex* | 0 | 1 | 2 | 3 | 4 | 5 | 6 | 7 |
|---|---|---|---|---|---|---|---|---|
| $x$ | 0 | 0 | 0 | 0 | 1 | 1 | 1 | 1 |
| $y$ | 0 | 0 | 1 | 1 | 0 | 0 | 1 | 1 |
| $z$ | 0 | 1 | 0 | 1 | 0 | 1 | 0 | 1 |

Let's repeat the three rotation transforms for rotating points about the $x$, $y$ and $z$ axes respectively, in their non-homogeneous form and substitute $c$ for cos and $s$ for sin to save space:

$$\text{rotate } \alpha \text{ about the } x \text{ axis} \quad \mathbf{R}_{\alpha,x} = \begin{bmatrix} 1 & 0 & 0 \\ 0 & c_\alpha & -s_\alpha \\ 0 & s_\alpha & c_\alpha \end{bmatrix}$$

$$\text{rotate } \beta \text{ about the } y \text{ axis} \quad \mathbf{R}_{\beta,y} = \begin{bmatrix} c_\beta & 0 & s_\beta \\ 0 & 1 & 0 \\ -s_\beta & 0 & c_\beta \end{bmatrix}$$

$$\text{rotate } \gamma \text{ about the } z \text{ axis} \quad \mathbf{R}_{\gamma,z} = \begin{bmatrix} c_\gamma & -s_\gamma & 0 \\ s_\gamma & c_\gamma & 0 \\ 0 & 0 & 1 \end{bmatrix}.$$

We can create a composite rotation by placing $\mathbf{R}_{\alpha,x}$, $\mathbf{R}_{\beta,y}$ and $\mathbf{R}_{\gamma,z}$ in any sequence. As an example, let's choose the sequence $\mathbf{R}_{\gamma,z}\mathbf{R}_{\beta,y}\mathbf{R}_{\alpha,x}$

$$\mathbf{R}_{\gamma,z}\mathbf{R}_{\beta,y}\mathbf{R}_{\alpha,x} = \begin{bmatrix} c_\gamma & -s_\gamma & 0 \\ s_\gamma & c_\gamma & 0 \\ 0 & 0 & 1 \end{bmatrix} \begin{bmatrix} c_\beta & 0 & s_\beta \\ 0 & 1 & 0 \\ -s_\beta & 0 & c_\beta \end{bmatrix} \begin{bmatrix} 1 & 0 & 0 \\ 0 & c_\alpha & -s_\alpha \\ 0 & s_\alpha & c_\alpha \end{bmatrix}. \tag{6.1}$$

Multiplying the three matrices in (6.1) together we obtain

$$\begin{bmatrix} c_\gamma c_\beta & c_\gamma s_\beta s_\alpha - s_\gamma c_\alpha & c_\gamma s_\beta c_\alpha + s_\gamma s_\alpha \\ s_\gamma c_\beta & s_\gamma s_\beta s_\alpha + c_\gamma c_\alpha & s_\gamma s_\beta c_\alpha - c_\gamma s_\alpha \\ -s_\beta & c_\beta s_\alpha & c_\beta c_\alpha \end{bmatrix}, \tag{6.2}$$

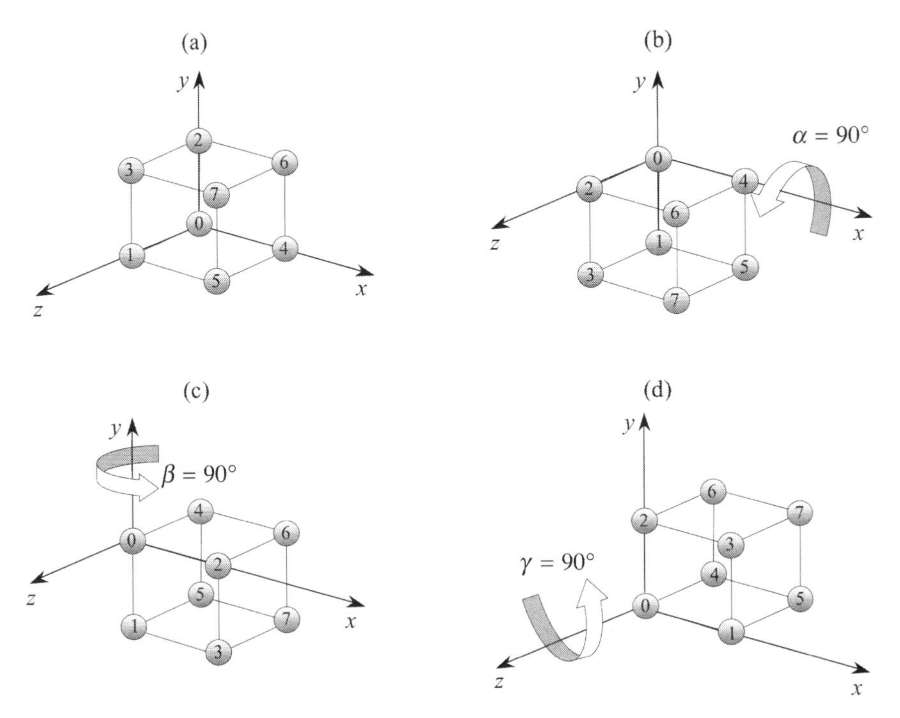


**Fig. 6.5** Four views of the unit cube before and during the three rotations

or using the more familiar notation:

$$\begin{bmatrix} \cos\gamma\cos\beta & \cos\gamma\sin\beta\sin\alpha - \sin\gamma\cos\alpha & \cos\gamma\sin\beta\cos\alpha + \sin\gamma\sin\alpha \\ \sin\gamma\cos\beta & \sin\gamma\sin\beta\sin\alpha + \cos\gamma\cos\alpha & \sin\gamma\sin\beta\cos\alpha - \cos\gamma\sin\alpha \\ -\sin\beta & \cos\beta\sin\alpha & \cos\beta\cos\alpha \end{bmatrix}.$$

Let's evaluate (6.2) by making $\alpha = \beta = \gamma = 90^\circ$:

$$\begin{bmatrix} 0 & 0 & 1 \\ 0 & 1 & 0 \\ -1 & 0 & 0 \end{bmatrix}. \tag{6.3}$$

The matrix (6.3) is equivalent to rotating a point 90° about the fixed $x$ axis, followed by a rotation of 90° about the fixed $y$ axis, followed by a rotation of 90° about the fixed $z$ axis. This rotation sequence is illustrated in Fig. 6.5.

Figure 6.5(a) shows the starting position of the cube; (b) shows its orientation after a 90° rotation about the $x$ axis; (c) shows its orientation after a further rotation of 90° about the $y$ axis; and (d) the cube's resting position after a rotation of 90° about the $z$ axis.

From Fig. 6.5(d) we see that the cube's coordinates are as shown in Table 6.2. We can confirm that these coordinates are correct by multiplying the cube's original coordinates shown in Table 6.1 by the matrix (6.3). Although it is not mathematically

Table 6.2 Vertex coordinates for the cube in Fig. 6.5(d)

| *vertex* | 0 | 1 | 2 | 3 | 4 | 5 | 6 | 7 |
|---|---|---|---|---|---|---|---|---|
| $x$ | 0 | 1 | 0 | 1 | 0 | 1 | 0 | 1 |
| $y$ | 0 | 0 | 1 | 1 | 0 | 0 | 1 | 1 |
| $z$ | 0 | 0 | 0 | 0 | −1 | −1 | −1 | −1 |

correct, we will show the matrix multiplying an array of coordinates as follows

$$\begin{bmatrix} 0 & 0 & 1 \\ 0 & 1 & 0 \\ -1 & 0 & 0 \end{bmatrix} \begin{bmatrix} 0 & 0 & 0 & 0 & 1 & 1 & 1 & 1 \\ 0 & 0 & 1 & 1 & 0 & 0 & 1 & 1 \\ 0 & 1 & 0 & 1 & 0 & 1 & 0 & 1 \end{bmatrix}$$

$$= \begin{bmatrix} 0 & 1 & 0 & 1 & 0 & 1 & 0 & 1 \\ 0 & 0 & 1 & 1 & 0 & 0 & 1 & 1 \\ 0 & 0 & 0 & 0 & -1 & -1 & -1 & -1 \end{bmatrix},$$

which agree with the coordinates in Table 6.2.

Naturally, any three angles can be chosen to rotate a point about the fixed axes, but it does become difficult to visualise without an interactive cgi system. Note that the determinant of (6.3) is 1, which is as expected.

An observation we made with 2D rotations is that they are additive: i.e. $\mathbf{R}_\alpha$ followed by $\mathbf{R}_\beta$ is equivalent to $\mathbf{R}_{\alpha+\beta}$. But something equally important is that rotations in 2D commute:

$$\mathbf{R}_\alpha \mathbf{R}_\beta = \mathbf{R}_\beta \mathbf{R}_\alpha = \mathbf{R}_{\alpha+\beta} = \mathbf{R}_{\beta+\alpha},$$

whereas, in general, 3D rotations are non-commutative. This is seen by considering a composite rotation formed by a rotation $\alpha$ about the $x$ axis $\mathbf{R}_{\alpha,x}$, followed by a rotation $\beta$ about the $z$ axis $\mathbf{R}_{\beta,z}$, and

$$\mathbf{R}_{\alpha,x} \mathbf{R}_{\beta,z} \neq \mathbf{R}_{\beta,z} \mathbf{R}_{\alpha,x}.$$

As an illustration, let's reverse the composite rotation computed above to $\mathbf{R}_{\alpha,x} \mathbf{R}_{\beta,y} \mathbf{R}_{\gamma,z}$:

$$\mathbf{R}_{\alpha,x} \mathbf{R}_{\beta,y} \mathbf{R}_{\gamma,z} = \begin{bmatrix} 1 & 0 & 0 \\ 0 & c_\alpha & -s_\alpha \\ 0 & s_\alpha & c_\alpha \end{bmatrix} \begin{bmatrix} c_\beta & 0 & s_\beta \\ 0 & 1 & 0 \\ -s_\beta & 0 & c_\beta \end{bmatrix} \begin{bmatrix} c_\gamma & -s_\gamma & 0 \\ s_\gamma & c_\gamma & 0 \\ 0 & 0 & 1 \end{bmatrix}. \tag{6.4}$$

Multiplying the three matrices in (6.4) together we obtain

$$\begin{bmatrix} c_\beta c_\gamma & -c_\beta s_\gamma & s_\beta \\ s_\alpha s_\beta c_\gamma + c_\alpha s_\gamma & -s_\alpha s_\beta s_\gamma + c_\alpha c_\gamma & -s_\alpha c_\beta \\ -c_\alpha s_\beta c_\gamma + s_\alpha s_\gamma & c_\alpha s_\beta s_\gamma + s_\alpha c_\gamma & c_\alpha c_\beta \end{bmatrix}, \tag{6.5}$$

or using the more familiar notation:

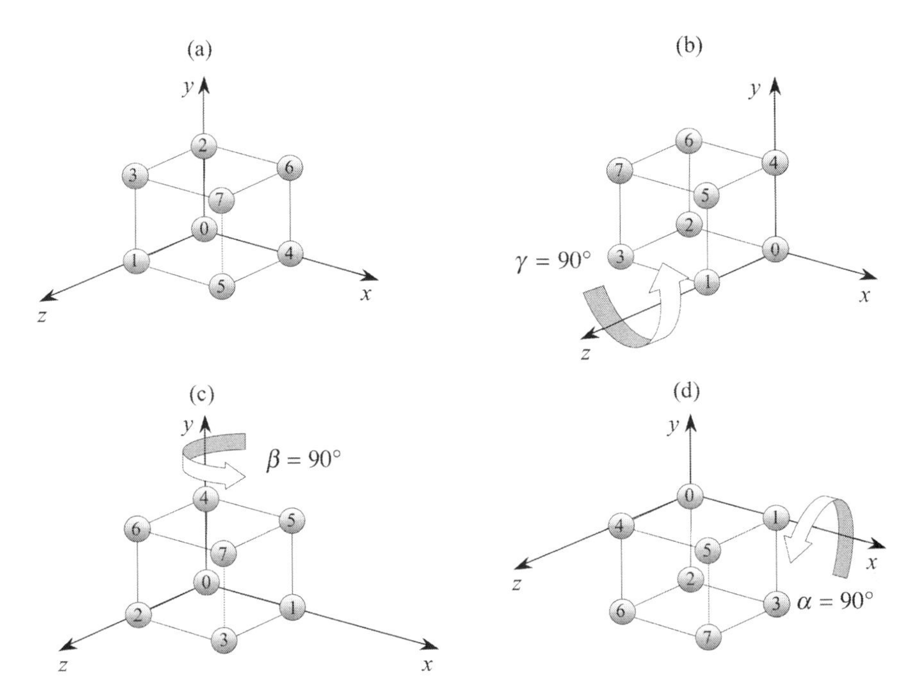


**Fig. 6.6** Four views of the unit cube using the rotation sequence $\mathbf{R}_{\alpha,x}\mathbf{R}_{\beta,y}\mathbf{R}_{\gamma,z}$

$$\begin{bmatrix} \cos\beta\cos\gamma & -\cos\beta\sin\gamma & \sin\beta \\ \sin\alpha\sin\beta\cos\gamma+\cos\alpha\sin\gamma & -\sin\alpha\sin\beta\sin\gamma+\cos\alpha\cos\gamma & -\sin\alpha\cos\beta \\ -\cos\alpha\sin\beta\cos\gamma+\sin\alpha\sin\gamma & \cos\alpha\sin\beta\sin\gamma+\sin\alpha\cos\gamma & \cos\alpha\cos\beta \end{bmatrix}.$$

Comparing (6.3) and (6.5) it can be seen that they are completely different.

Let's evaluate (6.5) by making $\alpha=\beta=\gamma=90^\circ$:

$$\begin{bmatrix} 0 & 0 & 1 \\ 0 & -1 & 0 \\ 1 & 0 & 0 \end{bmatrix}. \quad (6.6)$$

The matrix (6.6) is equivalent to rotating a point 90° about the fixed $z$ axis, followed by a rotation of 90° about the fixed $y$ axis, followed by a rotation of 90° about the fixed $x$ axis. This rotation sequence is illustrated in Fig. 6.6.

From Fig. 6.6(d) we see that the cube's coordinates are as shown in Table 6.3. We can confirm that these coordinates are correct by multiplying the cube's original coordinates shown in Table 6.1 by the matrix (6.6). We show the matrix multiplying an array of coordinates as before:

$$\begin{bmatrix} 0 & 0 & 1 \\ 0 & -1 & 0 \\ 1 & 0 & 0 \end{bmatrix}\begin{bmatrix} 0 & 0 & 0 & 0 & 1 & 1 & 1 & 1 \\ 0 & 0 & 1 & 1 & 0 & 0 & 1 & 1 \\ 0 & 1 & 0 & 1 & 0 & 1 & 0 & 1 \end{bmatrix}$$

**Table 6.3** Vertex coordinates for the cube in Fig. 6.6(d)

| *vertex* | 0 | 1 | 2 | 3 | 4 | 5 | 6 | 7 |
|---|---|---|---|---|---|---|---|---|
| $x$ | 0 | 1 | 0 | 1 | 0 | 1 | 0 | 1 |
| $y$ | 0 | 0 | −1 | −1 | 0 | 0 | −1 | −1 |
| $z$ | 0 | 0 | 0 | 0 | 1 | 1 | 1 | 1 |

$$= \begin{bmatrix} 0 & 1 & 0 & 1 & 0 & 1 & 0 & 1 \\ 0 & 0 & -1 & -1 & 0 & 0 & -1 & -1 \\ 0 & 0 & 0 & 0 & 1 & 1 & 1 & 1 \end{bmatrix},$$

which agree with the coordinates in Table 6.3, and we can safely conclude that, in general, 3D rotation transforms do not commute. Inspection of Fig. 6.6(d) shows that the unit cube has been rotated 180° about a vector $[1 \quad 0 \quad 1]^{\mathrm{T}}$.

So far we have created three composite rotations comprising individual rotations about the $x$, $y$ and $z$-axes: $\mathbf{R}_{\alpha,x}\mathbf{R}_{\beta,y}\mathbf{R}_{\gamma,z}$ and $\mathbf{R}_{\gamma,z}\mathbf{R}_{\beta,y}\mathbf{R}_{\alpha,x}$. But there is nothing stopping us from creating other combinations such as $\mathbf{R}_{\alpha,x}\mathbf{R}_{\beta,y}\mathbf{R}_{\gamma,x}$ or $\mathbf{R}_{\alpha,z}\mathbf{R}_{\beta,y}\mathbf{R}_{\gamma,z}$ that include two rotations about the same axis. In fact, there are twelve possible combinations:

$$\begin{array}{llll} \mathbf{R}_{\alpha,x}\mathbf{R}_{\beta,y}\mathbf{R}_{\gamma,x}, & \mathbf{R}_{\alpha,x}\mathbf{R}_{\beta,y}\mathbf{R}_{\gamma,z}, & \mathbf{R}_{\alpha,x}\mathbf{R}_{\beta,z}\mathbf{R}_{\gamma,x}, & \mathbf{R}_{\alpha,x}\mathbf{R}_{\beta,z}\mathbf{R}_{\gamma,y} \\ \mathbf{R}_{\alpha,y}\mathbf{R}_{\beta,x}\mathbf{R}_{\gamma,y}, & \mathbf{R}_{\alpha,y}\mathbf{R}_{\beta,x}\mathbf{R}_{\gamma,z}, & \mathbf{R}_{\alpha,y}\mathbf{R}_{\beta,z}\mathbf{R}_{\gamma,x}, & \mathbf{R}_{\alpha,y}\mathbf{R}_{\beta,z}\mathbf{R}_{\gamma,y} \\ \mathbf{R}_{\alpha,z}\mathbf{R}_{\beta,x}\mathbf{R}_{\gamma,y}, & \mathbf{R}_{\alpha,z}\mathbf{R}_{\beta,x}\mathbf{R}_{\gamma,z}, & \mathbf{R}_{\alpha,z}\mathbf{R}_{\beta,y}\mathbf{R}_{\gamma,x}, & \mathbf{R}_{\alpha,z}\mathbf{R}_{\beta,y}\mathbf{R}_{\gamma,z} \end{array}$$

which are covered in detail in Appendix.

Now let's explore the role eigenvectors and eigenvalues play in 3D rotations.

## 6.7 3D Eigenvectors

In Chap. 5 we examined the characteristic equation used to identify any eigenvectors associated with a matrix. The eigenvector $\mathbf{v}$ satisfies the relationship

$$\mathbf{Av} = \lambda\mathbf{v},$$

where $\lambda$ is a scaling factor.

In the context of a 3D rotation matrix, an eigenvector is a vector scaled by $\lambda$ but not rotated, which implies that it is the axis of rotation. To illustrate this, let's identify the eigenvector for the composite rotation (6.3) above:

$$\mathbf{R}_{90°,z}\mathbf{R}_{90°,y}\mathbf{R}_{90°,x} = \begin{bmatrix} 0 & 0 & 1 \\ 0 & 1 & 0 \\ -1 & 0 & 0 \end{bmatrix}.$$

Figure 6.5 shows the effect of this composite rotation, which is nothing more than a rotation of 90° about the $y$ axis. Therefore, we should be able to extract this information from the above matrix.

We begin by writing the characteristic equation for the matrix:

$$\begin{vmatrix} 0-\lambda & 0 & 1 \\ 0 & 1-\lambda & 0 \\ -1 & 0 & 0-\lambda \end{vmatrix} = 0. \tag{6.7}$$

Expanding (6.7) using the top row we have

$$\begin{aligned} -\lambda \begin{vmatrix} 1-\lambda & 0 \\ 0 & -\lambda \end{vmatrix} + 1 \begin{vmatrix} 0 & 1-\lambda \\ -1 & 0 \end{vmatrix} &= 0 \\ -\lambda(-\lambda+\lambda^2)+1-\lambda &= 0 \\ \lambda^2-\lambda^3+1-\lambda &= 0 \\ -\lambda^3+\lambda^2-\lambda+1 &= 0 \\ \lambda^3-\lambda^2+\lambda &= 1. \end{aligned}$$

When working with $3 \times 3$ matrices we always end up with a cubic in $\lambda$, for which there can be three types of solution:

1. One real and two complex conjugate solutions.
2. Three real solutions including the possibility of a double solution.
3. Three distinct real solutions.

It is clear that $\lambda = 1$ is one such real root, which satisfies our requirement for an eigenvalue. We could also show that the other two roots are complex conjugates.

Substituting $\lambda = 1$ in the original equations associated with (6.7) to reveal the eigenvector, we have

$$\begin{cases} -x + 0y + z = 0 \\ 0x + 0y + 0z = 0 \\ -x + 0y - z = 0. \end{cases}$$

It is obvious from the 1st and 3rd equations that $x = z = 0$. However, all three equations multiply the $y$ term by zero, which implies that the associated eigenvector is of the form $[0 \quad k \quad 0]^{\mathrm{T}}$, which is the $y$ axis, as anticipated. Now let's find the angle of rotation.

Using one of the above rotation matrices $\mathbf{R}_{\beta,y}$ and the trace operation:

$$\mathbf{R}_{\beta,y} = \begin{bmatrix} \cos\beta & 0 & \sin\beta \\ 0 & 1 & 0 \\ -\sin\beta & 0 & \cos\beta \end{bmatrix}$$

$$\mathrm{Tr}(\mathbf{R}_{\beta,y}) = 1 + 2\cos\beta$$

therefore,

$$\beta = \arccos\big((\mathrm{Tr}(\mathbf{R}_{\beta,y}) - 1)/2\big).$$

To illustrate this, let $\beta = 90°$:

$$\mathbf{R}_{90°,y} = \begin{bmatrix} 0 & 0 & 1 \\ 0 & 1 & 0 \\ -1 & 0 & 0 \end{bmatrix}$$

$$\mathrm{Tr}(\mathbf{R}_{90°,y}) = 1$$

therefore,

$$\beta = \arccos\big((1-1)/2\big) = 90°.$$

Let's choose another matrix and repeat the above:

$$\mathbf{R}_{\alpha,x} = \begin{bmatrix} 1 & 0 & 0 \\ 0 & \cos\alpha & -\sin\alpha \\ 0 & \sin\alpha & \cos\alpha \end{bmatrix}.$$

This time, let $\alpha = 45°$:

$$\mathbf{R}_{45°,x} = \begin{bmatrix} 1 & 0 & 0 \\ 0 & \sqrt{2}/2 & -\sqrt{2}/2 \\ 0 & \sqrt{2}/2 & \sqrt{2}/2 \end{bmatrix}$$

$$\mathrm{Tr}(\mathbf{R}_{45°,x}) = 1 + \sqrt{2}$$

therefore,

$$\alpha = \arccos\big((1+\sqrt{2}-1)/2\big) = 45°.$$

So we now have a mechanism to extract the axis and angle of rotation from a rotation matrix. However, the algorithm for identifying the axis is far from satisfactory, and later on we will discover that there is a similar technique which is readily programable.

For completeness, let's identify the axis and angle of rotation for the matrix (6.6):

$$\mathbf{R}_{90°,x}\mathbf{R}_{90°,y}\mathbf{R}_{90°,z} = \begin{bmatrix} 0 & 0 & 1 \\ 0 & -1 & 0 \\ 1 & 0 & 0 \end{bmatrix}.$$

Once more, we begin by writing the characteristic equation for the matrix:

$$\begin{vmatrix} 0-\lambda & 0 & 1 \\ 0 & -1-\lambda & 0 \\ 1 & 0 & 0-\lambda \end{vmatrix} = 0. \tag{6.8}$$

Expanding (6.8) using the top row we have

$$
\begin{aligned}
-\lambda \begin{vmatrix} -1-\lambda & 0 \\ 0 & -\lambda \end{vmatrix} + 1 \begin{vmatrix} 0 & -1-\lambda \\ 1 & 0 \end{vmatrix} &= 0 \\
-\lambda\left(-\lambda+\lambda^2\right) + 1 - \lambda &= 0 \\
\lambda^2 - \lambda^3 + 1 - \lambda &= 0 \\
-\lambda^3 + \lambda^2 - \lambda + 1 &= 0 \\
\lambda^3 - \lambda^2 + \lambda &= 1.
\end{aligned}
$$

Again, there is a single real root: $\lambda = 1$, and substituting this in the original equations associated with (6.8) to reveal the eigenvector, we have

$$
\begin{cases}
-x & + & 0y & + & z & = 0 \\
0x & - & 2y & + & 0z & = 0 \\
x & + & 0y & - & z & = 0.
\end{cases}
$$

It is obvious from the 1st and 3rd equations that $x = z$, and from the 2nd equation that $y = 0$, which implies that the associated eigenvector is of the form $[k \quad 0 \quad k]$, which is correct. Using the trace operation, we can write

$$
\mathrm{Tr}(\mathbf{R}_{90^\circ,x}\mathbf{R}_{90^\circ,y}\mathbf{R}_{90^\circ,z}) = -1,
$$

therefore,

$$
\beta = \arccos\left((-1-1)/2\right) = 180^\circ.
$$

As promised, let's explore another way of identifying the fixed axis of rotation, which is an eigenvector. Consider the following argument where $\mathbf{A}$ is a simple rotation transform:

If $\mathbf{v}$ is a fixed axis of rotation and $\mathbf{A}$ a rotation transform, then $\mathbf{v}$ suffers no rotation:

$$
\mathbf{A}\mathbf{v} = \mathbf{v}, \tag{6.9}
$$

similarly,

$$
\mathbf{A}^{\mathrm{T}}\mathbf{v} = \mathbf{v}. \tag{6.10}
$$

Subtracting (6.10) from (6.9), we have

$$
\mathbf{A}\mathbf{v} - \mathbf{A}^{\mathrm{T}}\mathbf{v} = \mathbf{0} \tag{6.11}
$$

$$
\left(\mathbf{A} - \mathbf{A}^{\mathrm{T}}\right)\mathbf{v} = \mathbf{0} \tag{6.12}
$$

where $\mathbf{0}$ is a null vector.

In Chap. 4 we defined an antisymmetric matrix $\mathbf{Q}$ as

$$\mathbf{Q} = \frac{1}{2}\left(\mathbf{A} - \mathbf{A}^{\mathrm{T}}\right), \tag{6.13}$$

therefore,

$$\left(\mathbf{A} - \mathbf{A}^{\mathrm{T}}\right) = 2\mathbf{Q}. \tag{6.14}$$

Substituting (6.14) in (6.12) we have

$$2\mathbf{Q}\mathbf{v} = \mathbf{0}$$
$$\mathbf{Q}\mathbf{v} = \mathbf{0}$$

which permits us to write

$$\begin{bmatrix} 0 & q_3 & -q_2 \\ -q_3 & 0 & q_1 \\ q_2 & -q_1 & 0 \end{bmatrix} \begin{bmatrix} v_1 \\ v_2 \\ v_3 \end{bmatrix} = \begin{bmatrix} 0 \\ 0 \\ 0 \end{bmatrix}, \tag{6.15}$$

where

$$q_1 = a_{23} - a_{32}$$
$$q_2 = a_{31} - a_{13}$$
$$q_3 = a_{12} - a_{21}.$$

Expanding (6.15) we have

$$0v_1 + q_3 v_2 - q_2 v_3 = 0$$
$$-q_3 v_1 + 0v_2 + q_1 v_3 = 0$$
$$q_2 v_1 - q_1 v_2 + 0v_3 = 0.$$

Obviously, one possible solution is $v_1 = v_2 = v_3 = 0$, but we seek a solution for $\mathbf{v}$ in terms of $q_1$, $q_2$ and $q_3$. A standard technique is to relax one of the $v$ terms, such as making $v_1 = 1$. Then

$$q_3 v_2 - q_2 v_3 = 0 \tag{6.16}$$
$$-q_3 + q_1 v_3 = 0 \tag{6.17}$$
$$q_2 - q_1 v_2 = 0. \tag{6.18}$$

From (6.18) we have

$$v_2 = \frac{q_2}{q_1}.$$

From (6.17) we have

$$v_3 = \frac{q_3}{q_1},$$

therefore, a solution is

$$\mathbf{v} = \begin{bmatrix} \frac{q_1}{q_1} & \frac{q_2}{q_1} & \frac{q_3}{q_1} \end{bmatrix}^{\mathrm{T}},$$

which in a non-homogeneous form is

$$\mathbf{v} = [q_1 \quad q_2 \quad q_3]^{\mathrm{T}},$$

or in terms of the original matrix:

$$\mathbf{v} = \big[(a_{23} - a_{32}) \quad (a_{31} - a_{13}) \quad (a_{12} - a_{21})\big]^{\mathrm{T}}, \tag{6.19}$$

which appears to be a rather elegant solution for finding the fixed axis of revolution.

Now let's put (6.19) to the test by recomputing the axis of rotation for the pure rotations $\mathbf{R}_{\alpha,x}$, $\mathbf{R}_{\beta,y}$ and $\mathbf{R}_{\gamma,z}$ where $\alpha = \beta = \gamma = 90°$.

$$\mathbf{R}_{90°,x} = \begin{bmatrix} 1 & 0 & 0 \\ 0 & 0 & -1 \\ 0 & 1 & 0 \end{bmatrix},$$

using (6.19) we have

$$\mathbf{v} = \big[(-1-1) \quad (0-0) \quad (0-0)\big] = [-2 \quad 0 \quad 0]^{\mathrm{T}},$$

which is the $x$ axis.

$$\mathbf{R}_{90°,y} = \begin{bmatrix} 0 & 0 & 1 \\ 0 & 1 & 0 \\ -1 & 0 & 0 \end{bmatrix},$$

using (6.19) we have

$$\mathbf{v} = \big[(0-0) \quad (-1-1) \quad (0-0)\big] = [0 \quad -2 \quad 0]^{\mathrm{T}},$$

which is the $y$ axis.

$$\mathbf{R}_{90°,z} = \begin{bmatrix} 0 & -1 & 0 \\ 1 & 0 & 0 \\ 0 & 0 & 1 \end{bmatrix},$$

using (6.19) we have

$$\mathbf{v} = \big[(0-0) \quad (0-0) \quad (-1-1)\big] = [0 \quad 0 \quad -2]^{\mathrm{T}},$$

which is the $z$ axis.

However, if we attempt to extract the axis of rotation from

$$\mathbf{R}_{90^\circ,x}\mathbf{R}_{90^\circ,y}\mathbf{R}_{90^\circ,z} = \begin{bmatrix} 0 & 0 & 1 \\ 0 & -1 & 0 \\ 1 & 0 & 0 \end{bmatrix},$$

we have a problem, because $q_1 = q_2 = q_3 = 0$. This is because $\mathbf{A} = \mathbf{A}^{\mathrm{T}}$ and the technique relies upon $\mathbf{A} \neq \mathbf{A}^{\mathrm{T}}$. So let's consider another approach based upon the fact that a rotation matrix always has a real eigenvalue $\lambda = 1$, which permits us to write

$$\mathbf{Av} = \lambda\mathbf{v}$$

$$\mathbf{Av} = \lambda\mathbf{Iv} = \mathbf{Iv}$$

$$(\mathbf{A} - \mathbf{I})\,\mathbf{v} = \mathbf{0}$$

therefore,

$$\begin{bmatrix} (a_{11} - 1) & a_{12} & a_{13} \\ a_{21} & (a_{22} - 1) & a_{23} \\ a_{31} & a_{32} & (a_{33} - 1) \end{bmatrix} \begin{bmatrix} v_1 \\ v_2 \\ v_3 \end{bmatrix} = \begin{bmatrix} 0 \\ 0 \\ 0 \end{bmatrix}. \tag{6.20}$$

Expanding (6.20) we have

$$(a_{11} - 1)v_1 + a_{12}v_2 + a_{13}v_3 = 0$$

$$a_{21}v_1 + (a_{22} - 1)v_2 + a_{23}v_3 = 0$$

$$a_{31}v_1 + a_{32}v_2 + (a_{33} - 1)v_3 = 0.$$

Once more, there exists a trivial solution where $v_1 = v_2 = v_3 = 0$, but to discover something more useful we can relax any one of the $v$ terms which gives us three equations in two unknowns. Let's make $v_1 = 0$:

$$a_{12}v_2 + a_{13}v_3 = -(a_{11} - 1) \tag{6.21}$$

$$(a_{22} - 1)v_2 + a_{23}v_3 = -a_{21} \tag{6.22}$$

$$a_{32}v_2 + (a_{33} - 1)v_3 = -a_{31}. \tag{6.23}$$

We are now faced with choosing a pair of equations to isolate $v_2$ and $v_3$. In fact, we have to consider all three pairings because it is possible that a future rotation matrix will contain a column with two zero elements, which could conflict with any pairing we make at this stage.

Let's begin by choosing (6.21) and (6.22). The solution employs the following strategy: Given the following matrix equation

$$\begin{bmatrix} a_1 & b_1 \\ a_2 & b_2 \end{bmatrix} \begin{bmatrix} x \\ y \end{bmatrix} = \begin{bmatrix} c_1 \\ c_2 \end{bmatrix},$$

then

$$\frac{x}{\begin{vmatrix} c_1 & b_1 \\ c_2 & b_2 \end{vmatrix}} = \frac{y}{\begin{vmatrix} a_1 & c_1 \\ a_2 & c_2 \end{vmatrix}} = \frac{1}{\begin{vmatrix} a_1 & b_1 \\ a_2 & b_2 \end{vmatrix}}.$$

Therefore, using the 1st and 2nd equations (6.21) and (6.22) we have

$$\frac{v_2}{\begin{vmatrix} -(a_{11}-1) & a_{13} \\ -a_{21} & a_{23} \end{vmatrix}} = \frac{v_3}{\begin{vmatrix} a_{12} & -(a_{11}-1) \\ (a_{22}-1) & -a_{21} \end{vmatrix}} = \frac{1}{\begin{vmatrix} a_{12} & a_{13} \\ (a_{22}-1) & a_{23} \end{vmatrix}}$$

$$v_1 = a_{12}a_{23} - a_{13}(a_{22} - 1)$$

$$v_2 = a_{13}a_{21} - a_{23}(a_{11} - 1)$$

$$v_3 = (a_{11} - 1)(a_{22} - 1) - a_{12}a_{21}.$$

Similarly, using the 1st and 3rd equations (6.21) and (6.23) we have

$$v_1 = a_{12}(a_{33} - 1) - a_{13}a_{32}$$

$$v_2 = a_{13}a_{31} - (a_{11} - 1)(a_{33} - 1)$$

$$v_3 = a_{32}(a_{11} - 1) - a_{12}a_{31}$$

and using the 2nd and 3rd equations (6.22) and (6.23) we have

$$v_1 = (a_{22} - 1)(a_{33} - 1) - a_{23}a_{32}$$

$$v_2 = a_{23}a_{31} - a_{21}(a_{33} - 1)$$

$$v_3 = a_{21}a_{32} - a_{31}(a_{22} - 1).$$

Now we have nine equations to cope with any eventuality. In fact, there is nothing to stop us from choosing any three that take our fancy, for example these three equations look interesting and sound:

$$v_1 = (a_{22} - 1)(a_{33} - 1) - a_{23}a_{32} \tag{6.24}$$

$$v_2 = (a_{33} - 1)(a_{11} - 1) - a_{31}a_{13} \tag{6.25}$$

$$v_3 = (a_{11} - 1)(a_{22} - 1) - a_{12}a_{21}. \tag{6.26}$$

Therefore, the solution for the eigenvector is $[v_1 \quad v_2 \quad v_3]^T$. Note that the sign of $v_2$ has been reversed to maintain symmetry.

Let's test (6.24), (6.25) and (6.26) with the transforms used above.

$$\mathbf{R}_{90°,x} = \begin{bmatrix} 1 & 0 & 0 \\ 0 & 0 & -1 \\ 0 & 1 & 0 \end{bmatrix} \left\{ \begin{array}{llcll} v_1 = & (-1)(-1) & - & (-1)\times 1 & = 2 \\ v_2 = & (-1)(0) & - & 0 \times 0 & = 0 \\ v_3 = & (0)(-1) & - & 0 \times 0 & = 0 \end{array} \right.$$

$$\mathbf{R}_{90^\circ,y} = \begin{bmatrix} 0 & 0 & 1 \\ 0 & 1 & 0 \\ -1 & 0 & 0 \end{bmatrix} \begin{cases} v_1 = & (0)(-1) & - & 0 \times 0 & = 0 \\ v_2 = & (-1)(-1) & - & (-1) \times 1 & = 2 \\ v_3 = & (-1)(0) & - & 0 \times 0 & = 0 \end{cases}$$

$$\mathbf{R}_{90^\circ,z} = \begin{bmatrix} 0 & -1 & 0 \\ 1 & 0 & 0 \\ 0 & 0 & 1 \end{bmatrix} \begin{cases} v_1 = & (-1)(0) & - & 0 \times 0 & = 0 \\ v_2 = & (0)(-1) & - & 0 \times 0 & = 0 \\ v_3 = & (-1)(-1) & - & (-1) \times 1 & = 2 \end{cases}$$

$$\mathbf{R}_{90^\circ,x}\mathbf{R}_{90^\circ,y}\mathbf{R}_{90^\circ,z} = \begin{bmatrix} 0 & 0 & 1 \\ 0 & -1 & 0 \\ 1 & 0 & 0 \end{bmatrix} \begin{cases} v_1 = & (-2)(-1) & - & 0 \times 0 & = 2 \\ v_2 = & (-1)(-1) & - & 1 \times 1 & = 0 \\ v_3 = & (-1)(-2) & - & 0 \times (-1) & = 2 \end{cases}$$

$$\mathbf{R}_{90^\circ,z}\mathbf{R}_{90^\circ,y}\mathbf{R}_{90^\circ,x} = \begin{bmatrix} 0 & 0 & 1 \\ 0 & 1 & 0 \\ -1 & 0 & 0 \end{bmatrix} \begin{cases} v_1 = & (0)(-1) & - & 0 \times 0 & = 0 \\ v_2 = & (-1)(-1) & - & (-1) \times 1 & = 2 \\ v_3 = & (-1)(0) & - & 0 \times 0 & = 0. \end{cases}$$

We can see why the resulting vectors have components of 2 by evaluating a normal rotation transform:

$$\mathbf{R}_{\alpha,x} = \begin{bmatrix} 1 & 0 & 0 \\ 0 & c_\alpha & -s_\alpha \\ 0 & s_\alpha & c_\alpha \end{bmatrix} \begin{cases} v_1 = & (c_\alpha - 1)(c_\alpha - 1) & - & (-s_\alpha) \times (s_\alpha) & = 2(1 - c_\alpha) \\ v_2 = & (c_\alpha - 1)(0) & - & 0 \times 0 & = 0 \\ v_3 = & (0)(c_\alpha - 1) & - & 0 \times 0 & = 0. \end{cases}$$

We can see that when $\alpha = 90^\circ$, $v_1 = 2$.

## 6.8 Gimbal Lock

There are two potential problems with all of the above composite transforms. The first is the difficulty visualising the orientation of an object subjected to several rotations; the second is that they all suffer from what is called *gimbal lock*. From a visualisation point of view, if we use the transform $\mathbf{R}_{\gamma,z}\mathbf{R}_{\beta,y}\mathbf{R}_{\alpha,x}$ to animate an object and change $\gamma$, $\beta$ and $\alpha$ over a period of frames, it can be very difficult to predict the final movement and adjust the angles to achieve a desired effect. Gimbal lock, on the other hand, is a weakness associated with Euler rotations when certain combinations of angles are used.

To understand this phenomenon, consider a simple gimbal which is a pivoted support that permits rotation about an axis, as shown in Fig. 6.7(a). If two gimbals are combined, as shown in Fig. 6.7(b), the inner cradle remains level with some reference plane as the assembly rolls and pitches. Such a combination has two degrees of rotational freedom. By adding a third gimbal so that the entire structure is free to rotate about a vertical axis, an extra degree of rotational freedom is introduced and is often used for mounting a camera on a tripod, as shown in Fig. 6.7(c).

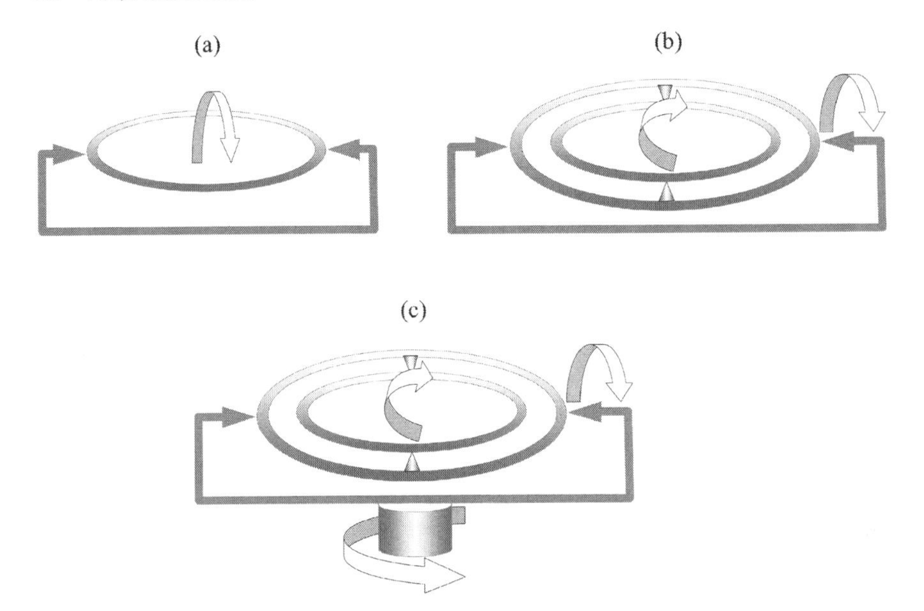


**Fig. 6.7** Three types of gimbal joints

A mechanical gimbal joint with three degrees of freedom is represented mathematically by a composite Euler rotation transform. For example, say we choose $\mathbf{R}_{90^\circ,y}\mathbf{R}_{90^\circ,x}\mathbf{R}_{90^\circ,z}$ to rotate our unit cube as shown in Fig. 6.8(a). The cube's faces containing vertices 1, 5, 7, 3 and 0, 2, 6, 4 are first rotated about the perpendicular $z$ axis, as shown in Fig. 6.8(b). The second transform rotates the cube's faces containing vertices 0, 4, 5, 1 and 2, 3, 7, 6 about the perpendicular $x$ axis, as shown in Fig. 6.8(c). If we now attempt to rotate the cube about the $y$ axis, as shown in Fig. 6.8(d), the cube's faces containing 0, 2, 6, 4 and 1, 5, 7, 3 are rotated again. Effectively we have lost the ability to rotate a cube about one of its axes, and such a condition is called *gimbal lock*. There is little we can do about this, apart from use another composite transform, but it, too, will have a similar restriction. For example, Appendix shows that $\mathbf{R}_{90^\circ,x}\mathbf{R}_{90^\circ,z}\mathbf{R}_{90^\circ,y}$, $\mathbf{R}_{90^\circ,y}\mathbf{R}_{90^\circ,z}\mathbf{R}_{90^\circ,x}$, $\mathbf{R}_{90^\circ,z}\mathbf{R}_{90^\circ,x}\mathbf{R}_{90^\circ,y}$ and $\mathbf{R}_{90^\circ,z}\mathbf{R}_{90^\circ,y}\mathbf{R}_{90^\circ,x}$ all possess a similar affliction. Fortunately, there are other ways of rotating an object, which we will explore later.

## 6.9 Yaw, Pitch and Roll

The above Euler rotations are also known as *yaw*, *pitch* and *roll*, and great care should be taken with these angles when referring to other books and technical papers. Sometimes a left-handed system of axes is used rather than a right-handed set, and the vertical axis may be the $y$ axis or the $z$ axis, and might even point downwards. Consequently, the matrices representing the rotations can vary greatly. In

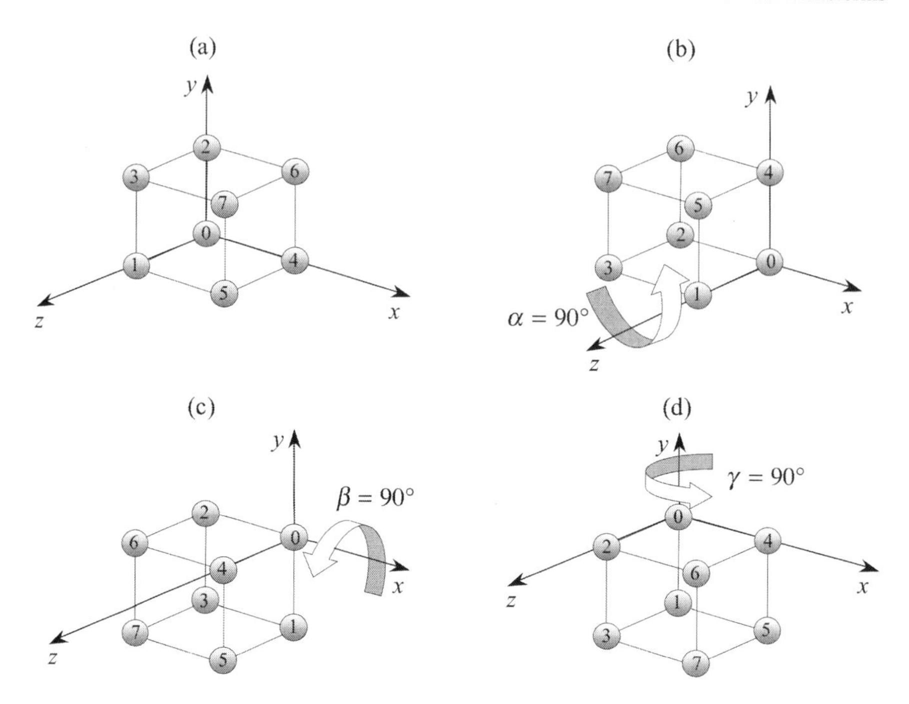


**Fig. 6.8** An example of gimbal lock

this text all Cartesian coordinate systems are right-handed, and the vertical axis is always the $y$ axis.

The terms yaw, pitch and roll are often used in aviation and to describe the motion of ships. For example, if a ship or aeroplane is heading in a particular direction, the axis aligned with the heading is the roll axis, as shown in Fig. 6.9(a). A perpendicular axis in the horizontal plane containing the heading axis is the pitch axis, as shown in Fig. 6.9(b). The axis perpendicular to both these axes is the yaw axis, as shown in Fig. 6.9(c). Clearly, there are many ways of aligning a set of Cartesian axes with the yaw, pitch and roll axes, and consequently, it is impossible to define an absolute set of yaw, pitch and roll transforms. However, if we choose the following alignment:

- the *roll* axis is the $z$ axis
- the *pitch* axis is the $x$ axis
- the *yaw* axis is the $y$ axis

we have the situation as shown in Fig. 6.10, and the transforms representing these rotations are as follows:

$$\mathbf{R}_{roll,z} = \begin{bmatrix} \cos roll & -\sin roll & 0 \\ \sin roll & \cos roll & 0 \\ 0 & 0 & 1 \end{bmatrix}$$

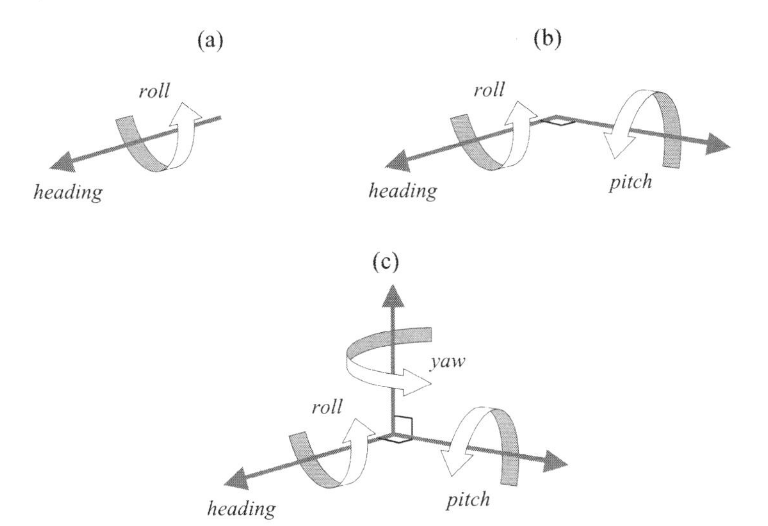


**Fig. 6.9** Definitions of yaw, pitch and roll

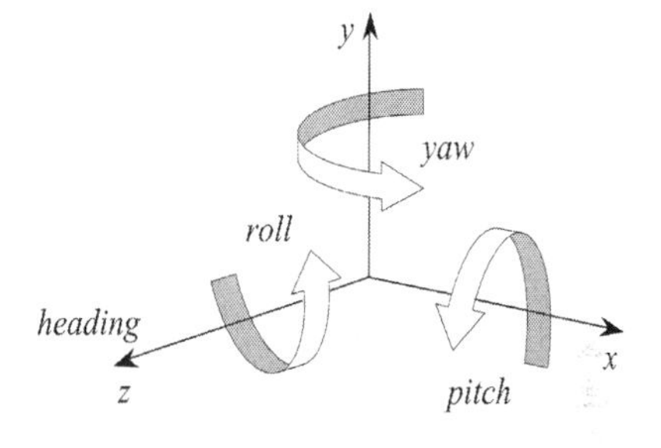


**Fig. 6.10** A convention for *roll*, *pitch* and *yaw* angles

$$\mathbf{R}_{pitch,x} = \begin{bmatrix} 1 & 0 & 0 \\ 0 & \cos pitch & -\sin pitch \\ 0 & \sin pitch & \cos pitch \end{bmatrix}$$

$$\mathbf{R}_{yaw,y} = \begin{bmatrix} \cos yaw & 0 & \sin yaw \\ 0 & 1 & 0 \\ -\sin yaw & 0 & \cos yaw \end{bmatrix}.$$

A common sequence for applying these rotations is *roll*, *pitch*, *yaw*, as seen in the following transform:

$$\begin{bmatrix} x' \\ y' \\ z' \end{bmatrix} = [\mathbf{R}_{yaw,y}][\mathbf{R}_{pitch,x}][\mathbf{R}_{roll,z}] \begin{bmatrix} x \\ y \\ z \end{bmatrix},$$

and if a translation is involved,

$$\begin{bmatrix} x' \\ y' \\ z' \\ 1 \end{bmatrix} = [\mathbf{T}_{d_x,d_y,d_z}][\mathbf{R}_{yaw,y}][\mathbf{R}_{pitch,x}][\mathbf{R}_{roll,z}] \begin{bmatrix} x \\ y \\ z \\ 1 \end{bmatrix}.$$

## 6.10 Rotation About an Arbitrary Axis

Now let's examine two ways of rotating a point about an arbitrary axis. The first technique uses matrices and trigonometry and is rather laborious.The second approach employs vector analysis and is quite succinct. Fortunately, they both arrive at the same result!

### *6.10.1 Matrices*

We begin by defining an axis using a unit vector $\hat{\mathbf{n}}$ about which a point $P$ is rotated $\alpha$ to $P'$ as shown in Fig. 6.11. And as we only have access to matrices that rotate points about the Cartesian axes, this unit vector has to be temporarily aligned with a Cartesian axis. In the following example we choose the $x$ axis. During the alignment process, the point $P$ is subjected to the transforms necessary to align the unit vector with the $x$ axis. We then rotate $P$, $\alpha$ about the $x$ axis. To complete the operation, the rotated point is subjected to the transforms that return the unit vector to its original position. Although matrices provide a powerful tool for undertaking this sort of work, it is nevertheless extremely tedious, but is a good exercise for improving one's algebraic skills!

Figure 6.11 shows a point $P(x, y, z)$ to be rotated through an angle $\alpha$ to $P'(x', y', z')$ about an axis defined by

$$\hat{\mathbf{n}} = a\mathbf{i} + b\mathbf{j} + c\mathbf{k}.$$

The transforms to achieve this operation can be expressed as follows:

$$\begin{bmatrix} x' \\ y' \\ z' \end{bmatrix} = [\mathbf{R}_{-\phi,y}][\mathbf{R}_{\theta,z}][\mathbf{R}_{\alpha,x}][\mathbf{R}_{-\theta,z}][\mathbf{R}_{\phi,y}] \begin{bmatrix} x \\ y \\ z \end{bmatrix},$$

which aligns the axis of rotation with the $x$ axis, performs the rotation of $P$ through an angle $\alpha$ about the $x$ axis, and returns the axis of rotation back to its original position. Therefore,

$$\mathbf{R}_{\phi,y} = \begin{bmatrix} \cos\phi & 0 & \sin\phi \\ 0 & 1 & 0 \\ -\sin\phi & 0 & \cos\phi \end{bmatrix}, \qquad \mathbf{R}_{-\theta,z} = \begin{bmatrix} \cos\theta & \sin\theta & 0 \\ -\sin\theta & \cos\theta & 0 \\ 0 & 0 & 1 \end{bmatrix}$$

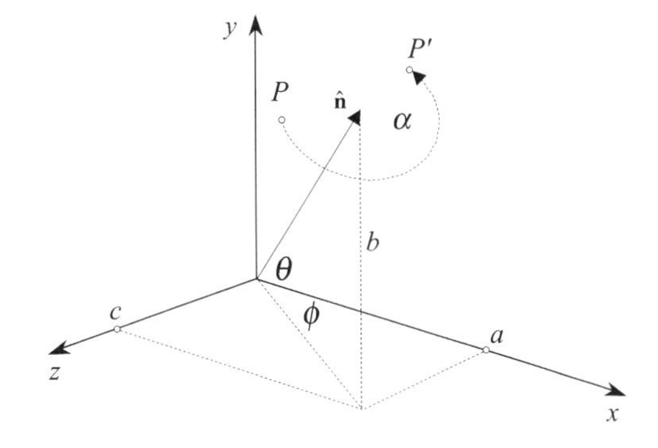


**Fig. 6.11** The geometry associated with rotating a point about an arbitrary axis

$$\mathbf{R}_{\alpha,x} = \begin{bmatrix} 1 & 0 & 0 \\ 0 & \cos\alpha & -\sin\alpha \\ 0 & \sin\alpha & \cos\alpha \end{bmatrix}, \qquad \mathbf{R}_{\theta,z} = \begin{bmatrix} \cos\theta & -\sin\theta & 0 \\ \sin\theta & \cos\theta & 0 \\ 0 & 0 & 1 \end{bmatrix}$$

$$\mathbf{R}_{-\phi,y} = \begin{bmatrix} \cos\phi & 0 & -\sin\phi \\ 0 & 1 & 0 \\ \sin\phi & 0 & \cos\phi \end{bmatrix}.$$

Let

$$\mathbf{R}_{-\phi,y}\mathbf{R}_{\theta,z}\mathbf{R}_{\alpha,x}\mathbf{R}_{-\theta,z}\mathbf{R}_{\phi,y} = \begin{bmatrix} a_{11} & a_{12} & a_{13} \\ a_{21} & a_{22} & a_{23} \\ a_{31} & a_{32} & a_{33} \end{bmatrix},$$

where by multiplying the matrices together we find that:

$$\begin{aligned}
a_{11} &= \cos^2\phi\cos^2\theta + \cos^2\phi\sin^2\theta\cos\alpha + \sin^2\phi\cos\alpha \\
a_{12} &= \cos\phi\cos\theta\sin\theta - \cos\phi\sin\theta\cos\theta\cos\alpha - \sin\phi\cos\theta\sin\alpha \\
a_{13} &= \cos\phi\sin\phi\cos^2\theta + \cos\phi\sin\phi\sin^2\theta\cos\alpha + \sin^2\phi\sin\theta\sin\alpha \\
&\quad + \cos^2\phi\sin\theta\sin\alpha - \cos\phi\sin\phi\cos\alpha \\
a_{21} &= \sin\theta\cos\theta\cos\phi - \cos\theta\sin\theta\cos\phi\cos\alpha + \cos\theta\sin\phi\sin\alpha \\
a_{22} &= \sin^2\theta + \cos^2\theta\cos\alpha \\
a_{23} &= \sin\theta\cos\theta\sin\phi - \cos\theta\sin\theta\sin\phi\cos\alpha - \cos\theta\cos\phi\sin\alpha \\
a_{31} &= \cos\phi\sin\phi\cos^2\theta + \cos\phi\sin\phi\sin^2\theta\cos\alpha - \cos^2\phi\sin\theta\sin\alpha \\
&\quad - \cos\phi\sin\phi\cos\alpha \\
a_{32} &= \sin\phi\cos\theta\sin\theta - \sin\phi\sin\theta\cos\theta\cos\alpha + \cos\phi\cos\theta\sin\alpha \\
a_{33} &= \sin^2\phi\cos^2\theta + \sin^2\phi\sin^2\theta\cos\alpha - \cos\phi\sin\phi\sin\theta\sin\alpha \\
&\quad + \cos\phi\sin\phi\sin\theta\sin\alpha + \cos^2\phi\cos\alpha.
\end{aligned}$$

From Fig. 6.11 we compute the sin and cos of $\theta$ and $\phi$ in terms of $a$, $b$ and $c$, and then compute their equivalent $\sin^2$ and $\cos^2$ values:

$$\begin{aligned}
\cos\theta &= \sqrt{1-b^2} && \Rightarrow && \cos^2\theta = 1-b^2 \\
\sin\theta &= b && \Rightarrow && \sin^2\theta = b^2 \\
\cos\phi &= a/\sqrt{1-b^2} && \Rightarrow && \cos^2\phi = a^2/(1-b^2) \\
\sin\phi &= c/\sqrt{1-b^2} && \Rightarrow && \sin^2\phi = c^2/(1-b^2).
\end{aligned}$$

To find $a_{11}$:

$$\begin{aligned}
a_{11} &= \cos^2\phi\cos^2\theta + \cos^2\phi\sin^2\theta\cos\alpha + \sin^2\phi\cos\alpha \\
&= a^2 + \frac{a^2b^2}{1-b^2}\cos\alpha + \frac{c^2}{1-b^2}\cos\alpha \\
&= a^2 + \left(\frac{c^2+a^2b^2}{1-b^2}\right)\cos\alpha
\end{aligned}$$

but

$$a^2+b^2+c^2 = 1 \Rightarrow c^2 = 1-a^2-b^2$$

$$\begin{aligned}
a_{11} &= a^2 + \left(\frac{1-a^2-b^2+a^2b^2}{1-b^2}\right)\cos\alpha \\
&= a^2 + \left(\frac{(1-a^2)(1-b^2)}{1-b^2}\right)\cos\alpha \\
&= a^2(1-\cos\alpha)+\cos\alpha.
\end{aligned}$$

Let

$$K = 1-\cos\alpha,$$

then

$$a_{11} = a^2K + \cos\alpha.$$

To find $a_{12}$:

$$\begin{aligned}
a_{12} &= \cos\phi\cos\theta\sin\theta - \cos\phi\sin\theta\cos\theta\cos\alpha - \sin\phi\cos\theta\sin\alpha \\
&= \frac{a}{\sqrt{1-b^2}}\sqrt{1-b^2}b - \frac{a}{\sqrt{1-b^2}}b\sqrt{1-b^2}\cos\alpha - \frac{c}{\sqrt{1-b^2}}\sqrt{1-b^2}\sin\alpha \\
&= ab - ab\cos\alpha - c\sin\alpha \\
&= ab(1-\cos\alpha) - c\sin\alpha \\
a_{12} &= abK - c\sin\alpha.
\end{aligned}$$

To find $a_{13}$:

$$\begin{aligned}
a_{13} &= \cos\phi\sin\phi\cos^2\theta + \cos\phi\sin\phi\sin^2\theta\cos\alpha + \sin^2\phi\sin\theta\sin\alpha \\
&\quad + \cos^2\phi\sin\theta\sin\alpha - \cos\phi\sin\phi\cos\alpha \\
&= \cos\phi\sin\phi\cos^2\theta + \cos\phi\sin\phi\sin^2\theta\cos\alpha + \sin\theta\sin\alpha - \cos\phi\sin\phi\cos\alpha \\
&= \frac{a}{\sqrt{1-b^2}}\frac{c}{\sqrt{1-b^2}}(1-b^2) + \frac{a}{\sqrt{1-b^2}}\frac{c}{\sqrt{1-b^2}}b^2\cos\alpha + b\sin\alpha \\
&\quad - \frac{a}{\sqrt{1-b^2}}\frac{c}{\sqrt{1-b^2}}\cos\alpha \\
&= ac + ac\frac{b^2}{(1-b^2)}\cos\alpha + b\sin\alpha - \frac{ac}{(1-b^2)}\cos\alpha \\
&= ac + ac\frac{(b^2-1)}{(1-b^2)}\cos\alpha + b\sin\alpha \\
&= ac(1-\cos\alpha) + b\sin\alpha \\
a_{13} &= acK + b\sin\alpha.
\end{aligned}$$

Using similar algebraic methods, we discover that

$$\begin{aligned}
a_{21} &= abK + c\sin\alpha \\
a_{22} &= b^2K + \cos\alpha \\
a_{23} &= bcK - a\sin\alpha \\
a_{31} &= acK - b\sin\alpha \\
a_{32} &= bcK + a\sin\alpha \\
a_{33} &= c^2K + \cos\alpha
\end{aligned}$$

and our original matrix transform becomes:

$$\begin{bmatrix} x'_p \\ y'_p \\ z'_p \end{bmatrix} = \begin{bmatrix} a^2K + \cos\alpha & abK - c\sin\alpha & acK + b\sin\alpha \\ abK + c\sin\alpha & b^2K + \cos\alpha & bcK - a\sin\alpha \\ acK - b\sin\alpha & bcK + a\sin\alpha & c^2K + \cos\alpha \end{bmatrix} \begin{bmatrix} x_p \\ y_p \\ z_p \end{bmatrix},$$

where

$$K = 1 - \cos\alpha.$$

### 6.10.2 Vectors

Now let's solve the same problem using vectors. Figure 6.12 shows a view of the geometry associated with the task at hand. For clarification, Fig. 6.13 shows a cross-section and a plan view of the geometry.

The axis of rotation is given by the unit vector:

$$\hat{\mathbf{n}} = a\mathbf{i} + b\mathbf{j} + c\mathbf{k}.$$

$P(x_p, y_p\ z_p)$ is the point to be rotated by angle $\alpha$ to $P'(x'_p, y'_p, z'_p)$.

$O$ is the origin, whilst $\mathbf{p}$ and $\mathbf{p}'$ are position vectors for $P$ and $P'$ respectively.

From Figs. 6.12 and 6.13:

$$\mathbf{p}' = \overrightarrow{ON} + \overrightarrow{NQ} + \overrightarrow{QP'}.$$

To find $\overrightarrow{ON}$:

$$|\mathbf{n}| = |\mathbf{p}| \cos\theta = \hat{\mathbf{n}} \cdot \mathbf{p},$$

therefore,

$$\overrightarrow{ON} = \mathbf{n} = \hat{\mathbf{n}}(\hat{\mathbf{n}} \cdot \mathbf{p}).$$

To find $\overrightarrow{NQ}$:

$$\overrightarrow{NQ} = \frac{NQ}{NP}\mathbf{r} = \frac{NQ}{NP'}\mathbf{r} = \cos\alpha\ \mathbf{r},$$

but

$$\mathbf{p} = \mathbf{n} + \mathbf{r} = \hat{\mathbf{n}}(\hat{\mathbf{n}} \cdot \mathbf{p}) + \mathbf{r},$$

therefore,

$$\mathbf{r} = \mathbf{p} - \hat{\mathbf{n}}(\hat{\mathbf{n}} \cdot \mathbf{p}),$$

and

$$\overrightarrow{NQ} = \big(\mathbf{p} - \hat{\mathbf{n}}(\hat{\mathbf{n}} \cdot \mathbf{p})\big)\cos\alpha.$$

To find $\overrightarrow{QP'}$:

Let

$$\hat{\mathbf{n}} \times \mathbf{p} = \mathbf{w},$$

where

$$|\mathbf{w}| = |\hat{\mathbf{n}}| \cdot |\mathbf{p}| \sin\theta = |\mathbf{p}| \sin\theta,$$

but

$$|\mathbf{r}| = |\mathbf{p}| \sin\theta,$$

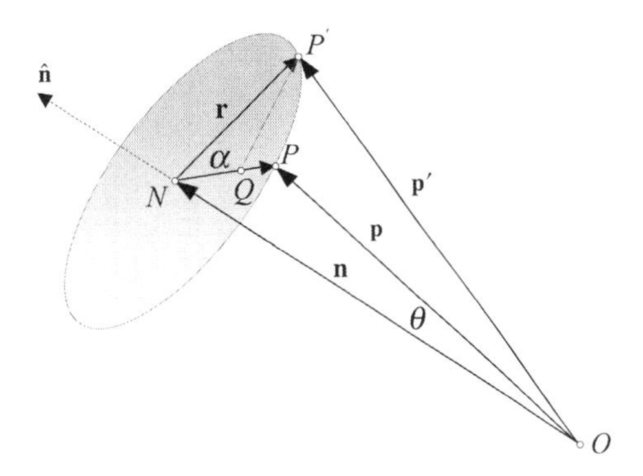


**Fig. 6.12** A view of the geometry associated with rotating a point about an arbitrary axis

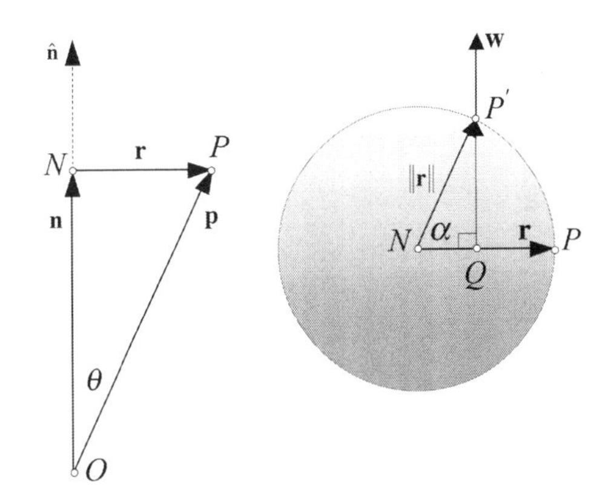


**Fig. 6.13** A cross-section and plan view of the geometry associated with rotating a point about an arbitrary axis

therefore,

$$|\mathbf{w}| = |\mathbf{r}|.$$

Now

$$\frac{QP'}{NP'} = \frac{QP'}{|\mathbf{r}|} = \frac{QP'}{|\mathbf{w}|} = \sin\alpha,$$

therefore,

$$\overrightarrow{QP'} = \mathbf{w}\sin\alpha = \hat{\mathbf{n}} \times \mathbf{p}\sin\alpha,$$

then

$$\mathbf{p}' = \hat{\mathbf{n}}(\hat{\mathbf{n}}\cdot\mathbf{p}) + \big(\mathbf{p} - \hat{\mathbf{n}}(\hat{\mathbf{n}}\cdot\mathbf{p})\big)\cos\alpha + \hat{\mathbf{n}} \times \mathbf{p}\sin\alpha,$$

and

$$\mathbf{p}' = \mathbf{p}\cos\alpha + \hat{\mathbf{n}}(\hat{\mathbf{n}}\cdot\mathbf{p})(1-\cos\alpha) + \hat{\mathbf{n}} \times \mathbf{p}\sin\alpha.$$

This is known as the Rodrigues rotation formula, as it was developed by the French mathematician, Olinde Rodrigues (1795–1851), who had also invented some of the ideas behind quaternions before Hamilton.

If we let

$$K = 1 - \cos\alpha,$$

then

$$\begin{aligned}
\mathbf{p}' &= \mathbf{p}\cos\alpha + \hat{\mathbf{n}}(\hat{\mathbf{n}}\cdot\mathbf{p})K + \hat{\mathbf{n}}\times\mathbf{p}\sin\alpha \\
&= (x_p\mathbf{i} + y_p\mathbf{j} + z_p\mathbf{k})\cos\alpha + (a\mathbf{i} + b\mathbf{j} + c\mathbf{k})(ax_p + by_p + cz_p)K \\
&\quad + \big((bz_p - cy_p)\mathbf{i} + (cx_p - az_p)j + (ay_p - bx_p)\mathbf{k}\big)\sin\alpha \\
&= \big(x_p\cos\alpha + a(ax_p + by_p + cz_p)K + (bz_p - cy_p)\sin\alpha\big)\mathbf{i} \\
&\quad + \big(y_p\cos\alpha + b(ax_p + by_p + cz_p)K + (cx_p - az_p)\sin\alpha\big)\mathbf{j} \\
&\quad + \big(z_p\cos\alpha + c(ax_p + by_p + cz_p)K + (ay_p - bx_p)\sin\alpha\big)\mathbf{k} \\
&= \big(x_p\big(a^2K + \cos\alpha\big) + y_p(abK - c\sin\alpha) + z_p(acK + b\sin\alpha)\big)\mathbf{i} \\
&\quad + \big(x_p(abK + c\sin\alpha) + y_p\big(b^2K + \cos\alpha\big) + z_p(bcK - a\sin\alpha)\big)\mathbf{j} \\
&\quad + \big(x_p(acK - b\sin\alpha) + y_p(bcK + a\sin\alpha) + z_p\big(c^2K + \cos\alpha\big)\big)\mathbf{k}
\end{aligned}$$

and the transform is:

$$\begin{bmatrix} x'_p \\ y'_p \\ z'_p \end{bmatrix} = \begin{bmatrix} a^2K + \cos\alpha & abK - c\sin\alpha & acK + b\sin\alpha \\ abK + c\sin\alpha & b^2K + \cos\alpha & bcK - a\sin\alpha \\ acK - b\sin\alpha & bcK + a\sin\alpha & c^2K + \cos\alpha \end{bmatrix} \begin{bmatrix} x_p \\ y_p \\ z_p \end{bmatrix},$$

which is identical to the transform derived using matrices.

Now let's test the transform with a simple example that can be easily verified. If we rotate the point $P(10, 0, 0)$, 180° about an axis defined by $\mathbf{n} = \mathbf{i} + \mathbf{j}$, it should end up at $P'(0, 10, 0)$.

Therefore

$$\alpha = 180^\circ, \qquad \cos\alpha = -1, \qquad \sin\alpha = 0, \qquad K = 2$$

$$a = \frac{\sqrt{2}}{2}, \qquad b = \frac{\sqrt{2}}{2}, \qquad c = 0$$

and

$$\begin{bmatrix} 0 \\ 10 \\ 0 \end{bmatrix} = \begin{bmatrix} 0 & 1 & 0 \\ 1 & 0 & 0 \\ 0 & 0 & 0 \end{bmatrix} \begin{bmatrix} 10 \\ 0 \\ 0 \end{bmatrix},$$

which is correct.

## 6.11 Worked Examples

*Example 1* Derive the matrix to scale by a factor of two relative to $(1, 2, 3)$, and show that the point $(1, 2, 3)$ is unmoved.

The transform is:

$$\mathbf{T}_{1,2,3}\mathbf{S}_{\times 2}\mathbf{T}_{-1,-2,-3}$$

$$\begin{aligned}
\begin{bmatrix} x' \\ y' \\ z' \\ 1 \end{bmatrix} &= \begin{bmatrix} 1 & 0 & 0 & 1 \\ 0 & 1 & 0 & 2 \\ 0 & 0 & 1 & 3 \\ 0 & 0 & 0 & 1 \end{bmatrix} \begin{bmatrix} 2 & 0 & 0 & 0 \\ 0 & 2 & 0 & 0 \\ 0 & 0 & 2 & 0 \\ 0 & 0 & 0 & 1 \end{bmatrix} \begin{bmatrix} 1 & 0 & 0 & -1 \\ 0 & 1 & 0 & -2 \\ 0 & 0 & 1 & -3 \\ 0 & 0 & 0 & 1 \end{bmatrix} \begin{bmatrix} x \\ y \\ z \\ 1 \end{bmatrix} \\
&= \begin{bmatrix} 1 & 0 & 0 & 1 \\ 0 & 1 & 0 & 2 \\ 0 & 0 & 1 & 3 \\ 0 & 0 & 0 & 1 \end{bmatrix} \begin{bmatrix} 2 & 0 & 0 & -2 \\ 0 & 2 & 0 & -4 \\ 0 & 0 & 2 & -6 \\ 0 & 0 & 0 & 1 \end{bmatrix} \begin{bmatrix} x \\ y \\ z \\ 1 \end{bmatrix} \\
&= \begin{bmatrix} 2 & 0 & 0 & -1 \\ 0 & 2 & 0 & -2 \\ 0 & 0 & 2 & -3 \\ 0 & 0 & 0 & 1 \end{bmatrix} \begin{bmatrix} x \\ y \\ z \\ 1 \end{bmatrix}.
\end{aligned}$$

Let's test it with the point $(1, 2, 3)$:

$$\begin{bmatrix} 1 \\ 2 \\ 3 \\ 1 \end{bmatrix} = \begin{bmatrix} 2 & 0 & 0 & -1 \\ 0 & 2 & 0 & -2 \\ 0 & 0 & 2 & -3 \\ 0 & 0 & 0 & 1 \end{bmatrix} \begin{bmatrix} 1 \\ 2 \\ 3 \\ 1 \end{bmatrix}$$

which confirms that it is unmoved.

*Example 2* Rotate the point $(1, 1, 0)$ 90° about the x axis. The matrix to rotate about the x axis is:

$$\begin{bmatrix} 1 & 0 & 0 \\ 0 & \cos\beta & -\sin\beta \\ 0 & \sin\beta & \cos\beta \end{bmatrix},$$

and when $\beta = 90°$:

$$\begin{bmatrix} 1 & 0 & 0 \\ 0 & 0 & -1 \\ 0 & 1 & 0 \end{bmatrix},$$

therefore, the transform is

$$\begin{bmatrix} 1 \\ 0 \\ 1 \end{bmatrix} = \begin{bmatrix} 1 & 0 & 0 \\ 0 & 0 & -1 \\ 0 & 1 & 0 \end{bmatrix} \begin{bmatrix} 1 \\ 1 \\ 0 \end{bmatrix},$$

which is correct.

*Example 3* Rotate the point $(1, 0, 0)$ $180^\circ$ about the vector $[1 \quad 0 \quad 1]^{\mathrm{T}}$.

The transform is

$$\begin{bmatrix} x'_p \\ y'_p \\ z'_p \end{bmatrix} = \begin{bmatrix} a^2K + \cos\alpha & abK - c\sin\alpha & acK + b\sin\alpha \\ abK + c\sin\alpha & b^2K + \cos\alpha & bcK - a\sin\alpha \\ acK - b\sin\alpha & bcK + a\sin\alpha & c^2K + \cos\alpha \end{bmatrix} \begin{bmatrix} x_p \\ y_p \\ z_p \end{bmatrix},$$

where

$$K = 1 - \cos\alpha,$$

and the unit vector is $[a \quad b \quad c]^{\mathrm{T}}$.

As $\alpha = 180^\circ$, $K = 2$ and $a = \frac{\sqrt{2}}{2}$, $b = 0$, $c = \frac{\sqrt{2}}{2}$.

Therefore,

$$\begin{bmatrix} 0 \\ 0 \\ 1 \end{bmatrix} = \begin{bmatrix} 0 & 0 & 1 \\ 0 & -1 & 0 \\ 1 & 0 & 0 \end{bmatrix} \begin{bmatrix} 1 \\ 0 \\ 0 \end{bmatrix},$$

which is correct.

*Example 4* Compute the eigenvector and eigenvalue from the transform in Example 3.

Using the above equations:

$$v_1 = (a_{22} - 1)(a_{33} - 1) - a_{23}a_{32}$$
$$v_2 = (a_{33} - 1)(a_{11} - 1) - a_{31}a_{13}$$
$$v_3 = (a_{11} - 1)(a_{22} - 1) - a_{12}a_{21}$$

where

$$\begin{bmatrix} a_{11} & a_{12} & a_{13} \\ a_{21} & a_{22} & a_{23} \\ a_{31} & a_{32} & a_{33} \end{bmatrix} = \begin{bmatrix} 0 & 0 & 1 \\ 0 & -1 & 0 \\ 1 & 0 & 0 \end{bmatrix},$$

then

$$v_1 = (-1 - 1)(0 - 1) = 2$$
$$v_2 = (0 - 1)(0 - 1) - 1 = 0$$
$$v_3 = (0 - 1)(-1 - 1) = 2$$

which is the vector $[2 \quad 0 \quad 2]^{\mathrm{T}}$.

To find the eigenvalue, we use

$$\mathrm{Tr}(\mathbf{R}) = 1 + 2\cos\beta$$
$$-1 = 1 + 2\cos\beta$$

$$
\begin{aligned}
-1 &= \cos\beta \\
180^\circ &= \beta.
\end{aligned}
$$

## 6.12 Summary

In this chapter we have examined various 3D transforms including: scale, translate, shear, reflection and rotate. We have seen that matrix notation provides an elegant way to express transforms and develop new ones. Out of all of the transforms, rotation is the most interesting as there are so many combinations to consider. In their simplest form, rotations are restricted to one of the three Cartesian axes, but by employing homogeneous coordinates, the translation transform can be used to rotate points about an off-set axis parallel with one of the Cartesian axes.

Composite Euler rotations are created by combining the matrices representing the individual rotations about three successive axes, for which there are twelve combinations. Unfortunately, one of the problems with such transforms is that they suffer from gimbal lock, where one degree of freedom is lost under certain angle combinations. Another problem, is that it is difficult to predict how a point moves in space when animated by a composite transform, although they are widely used for positioning objects in world space.

We have also seen how to compute the eigenvector associated with a rotation transform, and how it represents the axis about which rotation occurs, and the eigenvalue represents the angle of rotation.

Finally, matrices and vectors were used to develop a transform for rotating a point about an arbitrary axis.

In the next chapter we examine how a quaternion is expressed as a matrix, and how they can be used to rotate points about an axis.

# Chapter 7
# Quaternions

## 7.1 Introduction

Quaternions were invented by the Irish mathematician Sir William Rowan Hamilton (1805–1865) in 1843. Sir William was looking for a way to represent complex numbers in higher dimensions, and it took 15 years of toil before he stumbled upon the idea of using a 4D notation—hence the name 'quaternion'. Although a quaternion is a hyper-complex number, it does have a matrix form, which is derived in this chapter.

Knowing that a complex number is the combination of a real and imaginary quantity: $a + ib$, it is tempting to assume that its 3D equivalent is $a + ib + jc$ where $i^2 = j^2 = -1$. Unfortunately, when Hamilton formed the product of two such objects, he could not resolve the dyads $ij$ and $ji$, and went on to explore an extension $a + ib + jc + kd$ where $i^2 = j^2 = k^2 = -1$. This too, presented problems with the dyads $ij$, $jk$, $ki$ and their mirrors $ji$, $kj$ and $ik$. But after many years of thought Hamilton stumbled across the rules:

$$i^2 = j^2 = k^2 = ijk = -1$$

$$ij = k, \qquad jk = i, \qquad ki = j$$

$$ji = -k, \qquad kj = -i, \qquad ik = -j.$$

Although quaternions had some enthusiastic supporters, there were many mathematicians and scientists who were suspicious of the need to involve so many imaginary terms.

In 1881, the American mathematician Josiah Gibbs (1839–1903), at Yale University, resolved the problem by suggesting that the three imaginary quantities could be viewed as a 3D vector and changed the $ib + jc + kd$ into $b\mathbf{i} + c\mathbf{j} + d\mathbf{k}$, where $\mathbf{i}$, $\mathbf{j}$ and $\mathbf{k}$ are unit Cartesian vectors. Today, there are two ways of defining a quaternion:

$$q = [s, \mathbf{v}]$$

$$q = [s + \mathbf{v}].$$

J. Vince, *Matrix Transforms for Computer Games and Animation*,
DOI 10.1007/978-1-4471-4321-5_7, © Springer-Verlag London 2012

The difference is rather subtle: the first creates an ordered pair, and separates the scalar and the vector with a comma, whereas the second preserves the '+' sign as used in complex numbers. In this chapter we employ the comma as a separator.

It can be shown that quaternions can rotate points about an arbitrary axis, and hence the orientation of objects and a virtual camera. In order to develop the equation that performs this transformation we will have to understand the action of quaternions in the context of rotations.

A quaternion $q$ is the combination of a scalar and a vector:

$$q = [s, \mathbf{v}],$$

where $s$ is a scalar and $\mathbf{v}$ is a 3D vector. If we express the vector $\mathbf{v}$ in terms of its components, we have in an algebraic form

$$q = [s, x\mathbf{i} + y\mathbf{j} + z\mathbf{k}],$$

where $s$, $x$, $y$ and $z$ are real numbers.

## 7.2 Adding and Subtracting Quaternions

Given two quaternions $q_1$ and $q_2$:

$$q_1 = [s_1, \mathbf{v}_1] = [s_1, x_1\mathbf{i} + y_1\mathbf{j} + z_1\mathbf{k}]$$
$$q_2 = [s_2, \mathbf{v}_2] = [s_2, x_2\mathbf{i} + y_2\mathbf{j} + z_2\mathbf{k}]$$

they are equal if, and only if, their corresponding terms are equal. Furthermore, like vectors, they can be added and subtracted as follows:

$$q_1 \pm q_2 = \big[(s_1 \pm s_2), (x_1 \pm x_2)\mathbf{i} + (y_1 \pm y_2)\mathbf{j} + (z_1 \pm z_2)\mathbf{k}\big].$$

## 7.3 Multiplying Quaternions

When multiplying quaternions we must employ the following rules:

$$\mathbf{i}^2 = \mathbf{j}^2 = \mathbf{k}^2 = \mathbf{ijk} = -1$$
$$\mathbf{ij} = \mathbf{k}, \qquad \mathbf{jk} = \mathbf{i}, \qquad \mathbf{ki} = \mathbf{j}$$
$$\mathbf{ji} = -\mathbf{k}, \qquad \mathbf{kj} = -\mathbf{i}, \qquad \mathbf{ik} = -\mathbf{j}.$$

Note that whilst quaternion addition is commutative, the rules make quaternion products non-commutative.

Given two quaternions $q_1$ and $q_2$:

$$q_1 = [s_1, \mathbf{v}_1] = [s_1, x_1\mathbf{i} + y_1\mathbf{j} + z_1\mathbf{k}]$$

$$q_2 = [s_2, \mathbf{v}_2] = [s_2, x_2\mathbf{i} + y_2\mathbf{j} + z_2\mathbf{k}]$$

their product $q_1q_2$ is given by:

$$\begin{aligned} q_1q_2 = \big[&(s_1s_2 - x_1x_2 - y_1y_2 - z_1z_2), (s_1x_2 + s_2x_1 + y_1z_2 - y_2z_1)\mathbf{i} \\ &+ (s_1y_2 + s_2y_1 + z_1x_2 - z_2x_1)\mathbf{j} + (s_1z_2 + s_2z_1 + x_1y_2 - x_2y_1)\mathbf{k}\big] \end{aligned}$$

which can be rewritten using the dot and cross products as

$$q_1q_2 = [s_1s_2 - \mathbf{v}_1 \cdot \mathbf{v}_2, s_1\mathbf{v}_2 + s_2\mathbf{v}_1 + \mathbf{v}_1 \times \mathbf{v}_2],$$

where

$$s_1s_2 - \mathbf{v}_1 \cdot \mathbf{v}_2,$$

is a scalar, and

$$s_1\mathbf{v}_2 + s_2\mathbf{v}_1 + \mathbf{v}_1 \times \mathbf{v}_2,$$

is a vector.

## 7.4 Pure Quaternion

A pure quaternion has a zero scalar term:

$$q = [0, \mathbf{v}].$$

Therefore, given two pure quaternions:

$$q_1 = [0, \mathbf{v}_1] = [0, x_1\mathbf{i} + y_1\mathbf{j} + z_1\mathbf{k}]$$
$$q_2 = [0, \mathbf{v}_2] = [0, x_2\mathbf{i} + y_2\mathbf{j} + z_2\mathbf{k}]$$

their product is

$$q_1q_2 = [0, \mathbf{v}_1 \times \mathbf{v}_2],$$

which is another pure quaternion.

## 7.5 The Inverse Quaternion

Given the quaternion

$$q = [s, x\mathbf{i} + y\mathbf{j} + z\mathbf{k}],$$

its inverse $q^{-1}$ is given by

$$q^{-1} = \frac{[s, -x\mathbf{i} - y\mathbf{j} - z\mathbf{k}]}{|q|^2},$$

where $|q|$ is the magnitude, or modulus of $q$, and is equal to

$$|q| = \sqrt{s^2 + x^2 + y^2 + z^2}.$$

It can be shown that

$$qq^{-1} = q^{-1}q = 1.$$

## 7.6 Unit-Norm Quaternion

A unit-norm quaternion has a magnitude equal to 1:

$$|q| = \sqrt{s^2 + x^2 + y^2 + z^2} = 1.$$

## 7.7 Rotating Points About an Axis

Basically, quaternions are associated with vectors rather than individual points. Therefore, in order to manipulate a single vertex, it must be turned into a position vector, which has its tail at the origin. A vertex is then represented in quaternion form by its equivalent position vector with a zero scalar term. For example, a point $P(x, y, z)$ is represented in quaternion form by

$$p = [0, x\mathbf{i} + y\mathbf{j} + z\mathbf{k}],$$

which is transformed into another position vector using the process described below. The coordinates of the rotated point are the components of the rotated position vector. This may seem an indirect process, but in reality it turns out to be rather simple. Let's now consider how this is achieved.

It can be shown that a position vector $\mathbf{p}$ can be rotated through an angle $\theta$ about an axis using the following operation:

$$p' = qpq^{-1}, \tag{7.1}$$

where $p$ is a pure quaternion encoding the vector $\mathbf{p}$, $q$ is a unit-norm quaternion encoding the axis and angle of rotation, and $p'$ encodes the rotated vector $\mathbf{p}'$. (See the author's book *Quaternions for Computer Graphics*.) For example, to rotate a point $P(x, y, z)$ through an angle $\theta$ about an axis $\hat{\mathbf{u}}$, we use the following steps:

1. Convert the point $P(x, y, z)$ to a pure quaternion $p$:

$$p = [0, x\mathbf{i} + y\mathbf{j} + z\mathbf{k}].$$

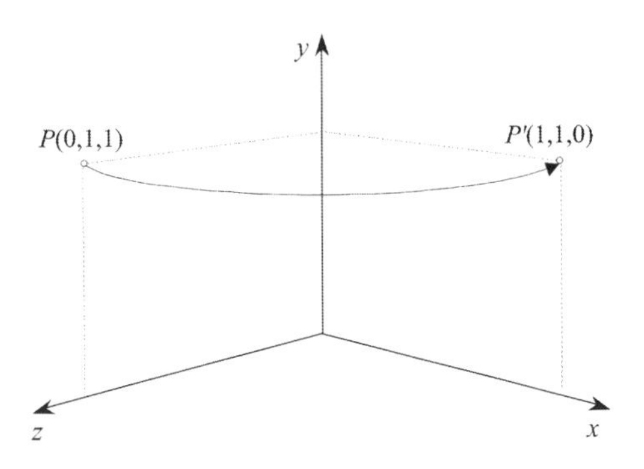


**Fig. 7.1** The point $P(0, 1, 1)$ is rotated to $P'(1, 1, 0)$ using a quaternion coincident with the $y$ axis

2. Define the axis of rotation as a unit vector $\hat{\mathbf{u}}$:

$$\hat{\mathbf{u}} = [x_u\mathbf{i} + y_u\mathbf{j} + z_u\mathbf{k}],$$

where

$$|\hat{\mathbf{u}}| = 1.$$

3. Define the transforming quaternion $q$, where the scalar term is $\cos\frac{\theta}{2}$ and the vector term is $\sin\frac{\theta}{2}\hat{\mathbf{u}}$:

$$q = \left[\cos\frac{\theta}{2}, \sin\frac{\theta}{2}\hat{\mathbf{u}}\right].$$

4. Define the inverse of the transforming quaternion $q^{-1}$:

$$q^{-1} = \left[\cos\frac{\theta}{2}, -\sin\frac{\theta}{2}\hat{\mathbf{u}}\right].$$

5. Compute $p'$:

$$p' = qpq^{-1}.$$

6. Unpack $(x', y', z')$ from $p'$:

$$P'(x', y', z') \quad \Leftarrow \quad p' = [0, x'\mathbf{i} + y'\mathbf{j} + z'\mathbf{k}].$$

We can verify the action of the above transform with a simple example. Consider the point $P(0, 1, 1)$ in Fig. 7.1 which is to be rotated 90° about the $y$ axis. We can see that the rotated point $P'$ has the coordinates $(1, 1, 0)$ which we will confirm algebraically. The point $P$ is represented by the quaternion $p$:

$$p = [0, 0\mathbf{i} + 1\mathbf{j} + 1\mathbf{k}],$$

and is rotated by evaluating the quaternion $p'$:

$$p' = qpq^{-1},$$

which will store the rotated coordinates. The axis of rotation is **j**, therefore the unit quaternion $q$ is given by

$$q = \left[\cos\frac{90^\circ}{2}, \sin\frac{90^\circ}{2}[0\mathbf{i} + \mathbf{j} + 0\mathbf{k}]\right]$$
$$= \left[\cos 45^\circ, 0\mathbf{i} + \sin 45^\circ\mathbf{j} + 0\mathbf{k}\right].$$

The inverse quaternion $q^{-1}$ is given by

$$q^{-1} = \frac{[\cos\frac{90^\circ}{2}, -\sin\frac{90^\circ}{2}[0\mathbf{i} + \mathbf{j} - 0\mathbf{k}]]}{|q|^2},$$

but as $q$ is a unit-norm quaternion, the denominator $|q|^2$ equals unity and can be ignored. Therefore

$$q^{-1} = \left[\cos 45^\circ, -0\mathbf{i} - \sin 45^\circ\mathbf{j} - 0\mathbf{k}\right].$$

Let's evaluate $qpq^{-1}$ in two stages: $(qp)q^{-1}$, and zero components will continue to be included for clarity.

1.

$$qp = \left[\cos 45^\circ, 0\mathbf{i} + \sin 45^\circ\mathbf{j} + 0\mathbf{k}\right][0 + 0\mathbf{i} + \mathbf{j} + 0\mathbf{k}]$$
$$= \left[-\sin 45^\circ, \sin 45^\circ\mathbf{i} + \cos 45^\circ\mathbf{j} + \cos 45^\circ\mathbf{k}\right].$$

2.

$$(qp)q^{-1} = \left[-\sin 45^\circ, \sin 45^\circ\mathbf{i} + \cos 45^\circ\mathbf{j} + \cos 45^\circ\mathbf{k}\right]$$
$$\times \left[\cos 45^\circ, -0\mathbf{i} - \sin 45^\circ\mathbf{j} - 0\mathbf{k}\right]$$
$$= \left[0, 2\cos 45^\circ \sin 45^\circ\mathbf{i} + \left(\cos^2 45^\circ + \sin^2 45^\circ\right)\mathbf{j}\right.$$
$$\left. + \left(\cos^2 45^\circ - \sin^2 45^\circ\right)\mathbf{k}\right]$$
$$p' = [0, \mathbf{i} + \mathbf{j} + 0\mathbf{k}]$$

and the vector component of $p'$ confirms that $P$ is indeed rotated to $(1, 1, 0)$.

We will evaluate one more example before continuing. Consider a rotation about the $z$ axis as illustrated in Fig. 7.2. The original point has coordinates $(0, 1, 1)$ and is rotated $-90^\circ$. From the figure we see that this should finish at $(1, 0, 1)$. This time the quaternion $q$ is defined by

$$q = \left[\cos\frac{-90^\circ}{2}, \sin\frac{-90^\circ}{2}[0\mathbf{i} + 0\mathbf{j} + \mathbf{k}]\right]$$
$$= \left[\cos 45^\circ, 0\mathbf{i} + 0\mathbf{j} - \sin 45^\circ\mathbf{k}\right]$$

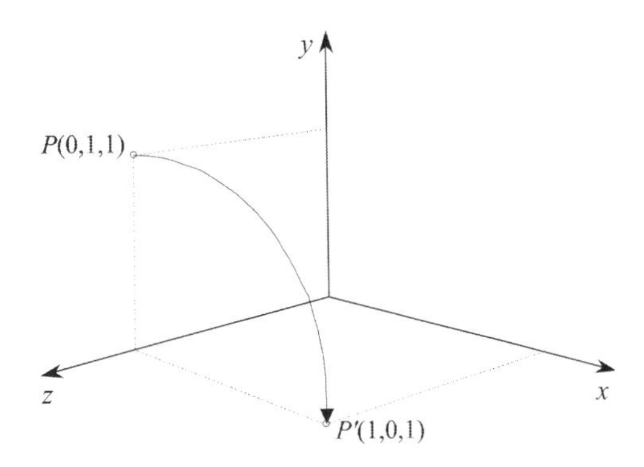


**Fig. 7.2** The point $P(0, 1, 1)$ is rotated $-90°$ to $P'(1, 0, 1)$ using a quaternion coincident with the $z$ axis

with its inverse

$$q^{-1} = \left[\cos 45°, 0\mathbf{i} + 0\mathbf{j} + \sin 45°\mathbf{k}\right],$$

and the point to be rotated in quaternion form is

$$p = [0, 0\mathbf{i} + \mathbf{i} + \mathbf{k}].$$

Evaluating this in two stages we have

1.

$$\begin{aligned} qp &= \left[\cos 45°, 0\mathbf{i} + 0\mathbf{j} - \sin 45°\mathbf{k}\right][0, 0\mathbf{i} + \mathbf{j} + \mathbf{k}] \\ &= \left[\sin 45°, \sin 45°\mathbf{i} + \cos 45°\mathbf{j} + \cos 45°\mathbf{k}\right]. \end{aligned}$$

2.

$$\begin{aligned} (qp)q^{-1} &= \left[\sin 45°, \sin 45°\mathbf{i} + \cos 45°\mathbf{j} + \cos 45°\mathbf{k}\right]\left[\cos 45°, 0\mathbf{i} + 0\mathbf{j} + \sin 45°\mathbf{k}\right] \\ &= \left[0, \sin 90°\mathbf{i} + \cos 90°\mathbf{j} + \mathbf{k}\right] \\ &= [0, \mathbf{i} + 0\mathbf{j} + \mathbf{k}]. \end{aligned}$$

The vector component of $p'$ confirms that $P$ is rotated to $(1, 0, 1)$.

## 7.8 Roll, Pitch and Yaw Quaternions

As described in the previous chapter, the rotational behaviour of planes and ships is often described in terms of roll, pitch and yaw. Roll is the rotation about the axis representing the direction of travel, say the $z$ axis, pitch is the rotation about the horizontal axis perpendicular to the $z$ axis, i.e. the $x$ axis, and yaw is the rotation about the vertical axis perpendicular to the $z$ axis, i.e. the $y$ axis. These are represented by the following quaternions:

$$q_{roll} = \left[\cos\frac{\theta}{2}, \sin\frac{\theta}{2}\mathbf{k}\right]$$

$$q_{pitch} = \left[\cos\frac{\theta}{2}, \sin\frac{\theta}{2}\mathbf{i}\right]$$

$$q_{yaw} = \left[\cos\frac{\theta}{2}, \sin\frac{\theta}{2}\mathbf{j}\right]$$

where $\theta$ is the angle of rotation.

These quaternions can be multiplied together to create a single quaternion representing a compound rotation. For example, if the quaternions are defined as

$$q_{roll} = \left[\cos\frac{roll}{2}, \sin\frac{roll}{2}\mathbf{k}\right]$$

$$q_{pitch} = \left[\cos\frac{pitch}{2}, \sin\frac{pitch}{2}\mathbf{i}\right]$$

$$q_{yaw} = \left[\cos\frac{yaw}{2}, \sin\frac{yaw}{2}\mathbf{j}\right]$$

they can be combined to a single quaternion $q$:

$$q = q_{yaw}q_{pitch}q_{roll} = [s, x\mathbf{i} + y\mathbf{j} + z\mathbf{k}],$$

where

$$s = \cos\frac{yaw}{2}\cos\frac{pitch}{2}\cos\frac{roll}{2} + \sin\frac{yaw}{2}\sin\frac{pitch}{2}\sin\frac{roll}{2}$$

$$x = \cos\frac{yaw}{2}\sin\frac{pitch}{2}\cos\frac{roll}{2} + \sin\frac{yaw}{2}\cos\frac{pitch}{2}\sin\frac{roll}{2}$$

$$y = \sin\frac{yaw}{2}\cos\frac{pitch}{2}\cos\frac{roll}{2} - \cos\frac{yaw}{2}\sin\frac{pitch}{2}\sin\frac{roll}{2}$$

$$z = \cos\frac{yaw}{2}\cos\frac{pitch}{2}\sin\frac{roll}{2} - \sin\frac{yaw}{2}\sin\frac{pitch}{2}\cos\frac{roll}{2}.$$

Let's examine this compound quaternion with an example. For instance, given the following conditions let's derive a single quaternion $q$ to represent the compound rotation:

$$roll = 90^\circ$$

$$pitch = 180^\circ$$

$$yaw = 0^\circ.$$

The values of $s, x, y, z$ are

$$s = 0$$

$$x = \cos 45^\circ$$

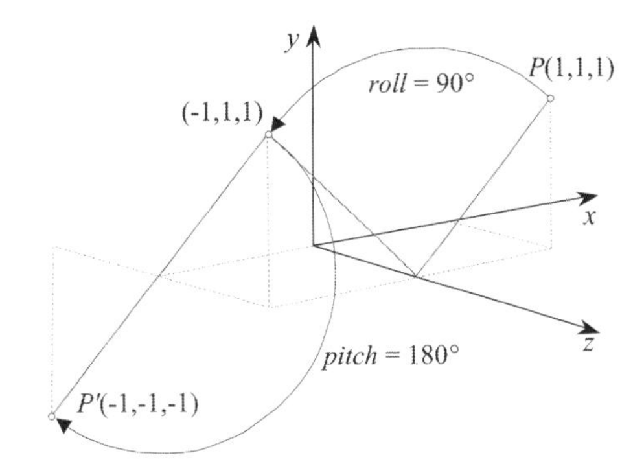


**Fig. 7.3** The point $P(1,1,1)$ is subject to a compound roll of 90° to $(-1,1,1)$ and a pitch of 180° and ends up at $P'(-1,-1,-1)$

$$y = -\sin 45^\circ$$
$$z = 0$$

and the quaternion $q$ is

$$q = \left[0, \cos 45^\circ \mathbf{i} - \sin 45^\circ \mathbf{j}\right].$$

If the point $P(1,1,1)$ is subjected to this compound rotation, the rotated point is computed using the standard quaternion transform:

$$p' = qpq^{-1}.$$

Let's evaluate $qpq^{-1}$ in two stages:

1.

$$\begin{aligned} qp &= \left[0, \cos 45^\circ \mathbf{i} - \sin 45^\circ \mathbf{j}\right][0, \mathbf{i} + \mathbf{j} + \mathbf{k}] \\ &= \left[0, -\sin 45^\circ \mathbf{i} - \cos 45^\circ \mathbf{j} + \left(\sin 45^\circ + \cos 45^\circ\right)\mathbf{k}\right]. \end{aligned}$$

2.

$$\begin{aligned} (qp)q^{-1} &= \left[0, -\sin 45^\circ \mathbf{i} - \cos 45^\circ \mathbf{j} + \left(\sin 45^\circ + \cos 45^\circ\right)\mathbf{k}\right] \\ &\quad \times \left[0, -\cos 45^\circ \mathbf{i} + \sin 45^\circ \mathbf{j}\right] \\ p' &= [0, -\mathbf{i} - \mathbf{j} - \mathbf{k}]. \end{aligned}$$

Therefore, the coordinates of the rotated point are $(-1,-1,-1)$ which can be confirmed from Fig. 7.3.

## 7.9 Quaternions in Matrix Form

Having discovered a vector equation to represent the triple $qpq^{-1}$, let's continue and convert it into a matrix. We will explore two methods: The first is a simple vectorial method which translates the vector equation representing $qpq^{-1}$ directly into matrix form; the second method uses matrix algebra to develop a rather cunning solution.

### 7.9.1 Vector Method

For the vector method it is convenient to describe the unit-norm quaternion as

$$\begin{aligned} q &= [s, \mathbf{v}] \\ &= [s, x\mathbf{i} + y\mathbf{j} + z\mathbf{k}] \end{aligned}$$

where

$$s^2 + |\mathbf{v}|^2 = 1,$$

and the pure quaternion as

$$\begin{aligned} p &= [0, \mathbf{p}] \\ &= [0, x_p\mathbf{i} + y_p\mathbf{j} + z_p\mathbf{k}]. \end{aligned}$$

A simple way to compute $qpq^{-1}$ is to use (7.1) and substitute $|\mathbf{v}|$ for $\lambda$:

$$\begin{aligned} qpq^{-1} &= \big[0, 2\lambda^2(\hat{\mathbf{v}} \cdot \mathbf{p})\hat{\mathbf{v}} + \big(s^2 - \lambda^2\big)\mathbf{p} + 2\lambda s\hat{\mathbf{v}} \times \mathbf{p}\big] \\ &= \big[0, 2|\mathbf{v}|^2(\hat{\mathbf{v}} \cdot \mathbf{p})\hat{\mathbf{v}} + \big(s^2 - |\mathbf{v}|^2\big)\mathbf{p} + 2|\mathbf{v}|s\hat{\mathbf{v}} \times \mathbf{p}\big]. \end{aligned}$$

Next, we substitute $\mathbf{v}$ for $|\mathbf{v}|\hat{\mathbf{v}}$:

$$qpq^{-1} = \big[0, 2(\mathbf{v} \cdot \mathbf{p})\mathbf{v} + \big(s^2 - |\mathbf{v}|^2\big)\mathbf{p} + 2s\mathbf{v} \times \mathbf{p}\big].$$

Finally, as we are working with unit-norm quaternions to prevent scaling

$$s^2 + |\mathbf{v}|^2 = 1,$$

and

$$s^2 - |\mathbf{v}|^2 = 2s^2 - 1,$$

therefore,

$$qpq^{-1} = \big[0, 2(\mathbf{v} \cdot \mathbf{p})\mathbf{v} + \big(2s^2 - 1\big)\mathbf{p} + 2s\mathbf{v} \times \mathbf{p}\big].$$

If we let $p' = qpq^{-1}$, which is a pure quaternion, we have

$$\begin{aligned} p' &= qpq^{-1} \\ &= \big[0, \mathbf{p}'\big] \end{aligned}$$

$$= \left[0, 2(\mathbf{v}\cdot\mathbf{p})\mathbf{v} + \left(2s^2 - 1\right)\mathbf{p} + 2s\mathbf{v}\times\mathbf{p}\right]$$

$$\mathbf{p}' = 2(\mathbf{v}\cdot\mathbf{p})\mathbf{v} + \left(2s^2 - 1\right)\mathbf{p} + 2s\mathbf{v}\times\mathbf{p}.$$

We are only interested in the rotated vector $\mathbf{p}'$ comprising the three terms $2(\mathbf{v}\cdot\mathbf{p})\mathbf{v}$, $(2s^2-1)\mathbf{p}$ and $2s\mathbf{v}\times\mathbf{p}$, which can be represented by three individual matrices and summed together.

$$\begin{aligned}
2(\mathbf{v}\cdot\mathbf{p})\mathbf{v} &= 2(xx_p + yy_p + zz_p)(x\mathbf{i} + y\mathbf{j} + z\mathbf{k}) \\
&= \begin{bmatrix} 2x^2 & 2xy & 2xz \\ 2xy & 2y^2 & 2yz \\ 2xz & 2yz & 2z^2 \end{bmatrix} \begin{bmatrix} x_p \\ y_p \\ z_p \end{bmatrix} \\
\left(2s^2-1\right)\mathbf{p} &= \left(2s^2-1\right)x_p\mathbf{i} + \left(2s^2-1\right)y_p\mathbf{j} + \left(2s^2-1\right)z_p\mathbf{k} \\
&= \begin{bmatrix} 2s^2-1 & 0 & 0 \\ 0 & 2s^2-1 & 0 \\ 0 & 0 & 2s^2-1 \end{bmatrix} \begin{bmatrix} x_p \\ y_p \\ z_p \end{bmatrix} \\
2s\mathbf{v}\times\mathbf{p} &= 2s\left((yz_p - zy_p)\mathbf{i} + (zx_p - xz_p)\mathbf{j} + (xy_p - yx_p)\mathbf{k}\right) \\
&= \begin{bmatrix} 0 & -2sz & 2sy \\ 2sz & 0 & -2sx \\ -2sy & 2sx & 0 \end{bmatrix} \begin{bmatrix} x_p \\ y_p \\ z_p \end{bmatrix}.
\end{aligned}$$

Adding these matrices together:

$$\mathbf{p}' = \begin{bmatrix} 2(s^2+x^2)-1 & 2(xy-sz) & 2(xz+sy) \\ 2(xy+sz) & 2(s^2+y^2)-1 & 2(yz-sx) \\ 2(xz-sy) & 2(yz+sx) & 2(s^2+z^2)-1 \end{bmatrix} \begin{bmatrix} x_p \\ y_p \\ z_p \end{bmatrix} \tag{7.2}$$

or

$$\mathbf{p}' = \begin{bmatrix} 1-2(y^2+z^2) & 2(xy-sz) & 2(xz+sy) \\ 2(xy+sz) & 1-2(x^2+z^2) & 2(yz-sx) \\ 2(xz-sy) & 2(yz+sx) & 1-2(x^2+y^2) \end{bmatrix} \begin{bmatrix} x_p \\ y_p \\ z_p \end{bmatrix} \tag{7.3}$$

where

$$\left[0, \mathbf{p}'\right] = qpq^{-1}.$$

Now let's reverse the product. To compute the vector part of $q^{-1}pq$ all that we have to do is reverse the sign of $2s\mathbf{v}\times\mathbf{p}$:

$$\mathbf{p}' = \begin{bmatrix} 2(s^2+x^2)-1 & 2(xy+sz) & 2(xz-sy) \\ 2(xy-sz) & 2(s^2+y^2)-1 & 2(yz+sx) \\ 2(xz+sy) & 2(yz-sx) & 2(s^2+z^2)-1 \end{bmatrix} \begin{bmatrix} x_p \\ y_p \\ z_p \end{bmatrix} \tag{7.4}$$

or

$$\mathbf{p}' = \begin{bmatrix} 1-2(y^2+z^2) & 2(xy+sz) & 2(xz-sy) \\ 2(xy-sz) & 1-2(x^2+z^2) & 2(yz+sx) \\ 2(xz+sy) & 2(yz-sx) & 1-2(x^2+y^2) \end{bmatrix} \begin{bmatrix} x_p \\ y_p \\ z_p \end{bmatrix} \tag{7.5}$$

where

$$[0, \mathbf{p}'] = q^{-1}pq.$$

Observe that (7.4) is the transpose of (7.2), and (7.5) is the transpose of (7.3).

### 7.9.2 Matrix Method

The second method to derive (7.1) employs the matrix representing a quaternion product:

$$q_a = [s_a, x_a\mathbf{i} + y_a\mathbf{j} + z_a\mathbf{k}]$$
$$q_b = [s_b, x_b\mathbf{i} + y_b\mathbf{j} + z_b\mathbf{k}]$$

and their product is

$$\begin{aligned} q_a q_b &= [s_a, x_a\mathbf{i} + y_a\mathbf{j} + z_a\mathbf{k}][s_b, x_b\mathbf{i} + y_b\mathbf{j} + z_b\mathbf{k}] \\ &= \big[s_a s_b - x_a x_b - y_a y_b - z_a z_b, \\ &\quad s_a(x_b\mathbf{i} + y_b\mathbf{j} + z_b\mathbf{k}) + s_b(x_a\mathbf{i} + y_a\mathbf{j} + z_a\mathbf{k}) \\ &\quad + (y_a z_b - y_b z_a)\mathbf{i} + (x_b z_a - x_a z_b)\mathbf{j} + (x_a y_b - x_b y_a)\mathbf{k}\big] \\ &= \big[s_a s_b - x_a x_b - y_a y_b - z_a z_b, \\ &\quad (s_a x_b + s_b x_a + y_a z_b - y_b z_a)\mathbf{i} \\ &\quad + (s_a y_b + s_b y_a + x_b z_a - x_a z_b)\mathbf{j} \\ &\quad + (s_a z_b + s_b z_a + x_a y_b - x_b y_a)\mathbf{k}\big] \\ &= \begin{bmatrix} s_a & -x_a & -y_a & -z_a \\ x_a & s_a & -z_a & y_a \\ y_a & z_a & s_a & -x_a \\ z_a & -y_a & x_a & s_a \end{bmatrix} \begin{bmatrix} s_b \\ x_b \\ y_b \\ z_b \end{bmatrix} = \mathbf{A}q_b. \end{aligned}$$

At this stage we have quaternion $q_a$ represented by matrix $\mathbf{A}$, and quaternion $q_b$ represented as a column vector. Now let's reverse the scenario without altering the result by making $q_b$ the matrix and $q_a$ the column vector:

$$q_a q_b = \begin{bmatrix} s_b & -x_b & -y_b & -z_b \\ x_b & s_b & z_b & -y_b \\ y_b & -z_b & s_b & x_b \\ z_b & y_b & -x_b & s_b \end{bmatrix} \begin{bmatrix} s_a \\ x_a \\ y_a \\ z_a \end{bmatrix} = \mathbf{B} q_a.$$

So now we have two ways of computing $q_a q_b$ and we need a way of distinguishing between the two matrices. Let **L** be the matrix that preserves the left-to-right quaternion sequence, and **R** be the matrix that reverses the sequence to right-to-left:

$$q_a q_b = \mathbf{L}(q_a) q_b = \begin{bmatrix} s_a & -x_a & -y_a & -z_a \\ x_a & s_a & -z_a & y_a \\ y_a & z_a & s_a & -x_a \\ z_a & -y_a & x_a & s_a \end{bmatrix} \begin{bmatrix} s_b \\ x_b \\ y_b \\ z_b \end{bmatrix}$$

$$q_a q_b = \mathbf{R}(q_b) q_a = \begin{bmatrix} s_b & -x_b & -y_b & -z_b \\ x_b & s_b & z_b & -y_b \\ y_b & -z_b & s_b & x_b \\ z_b & y_b & -x_b & s_b \end{bmatrix} \begin{bmatrix} s_a \\ x_a \\ y_a \\ z_a \end{bmatrix}.$$

Remember that $\mathbf{L}(q_a) q_b = \mathbf{R}(q_b) q_a$, as this is central to understanding the next stage. Furthermore, don't be surprised if you can't follow the argument in the first reading.

First, let's employ the matrices **L** and **R** to rearrange the quaternion product $q_a q_c q_b$ to $q_a q_b q_c$. i.e. move $q_c$ from the middle to the right-hand-side. We start with the quaternion product $q_a q_c q_b$ and divide it into two parts, $q_a q_c$ and $q_b$. We can do this because quaternion algebra is associative:

$$q_a q_c q_b = (q_a q_c) q_b.$$

We have already demonstrated above that the product $q_a q_c$ can be replaced by $\mathbf{L}(q_a) q_c$:

$$q_a q_c q_b = \mathbf{L}(q_a) q_c q_b.$$

We now have another two parts: $\mathbf{L}(q_a) q_c$ and $q_b$ which can be reversed using **R** without disturbing the result:

$$q_a q_c q_b = \mathbf{L}(q_a) q_c q_b = \mathbf{R}(q_b) \mathbf{L}(q_a) q_c,$$

which has achieved our objective to move $q_c$ to the right-hand-side. But the most important result is that the matrices $\mathbf{R}(q_b)$ and $\mathbf{L}(q_a)$ can be multiplied together to form a single matrix, which operates on $q_c$.

Now let's repeat the same process to rearrange the product $qpq^{-1}$. The objective is to move $p$ from the middle of $q$ and $q^{-1}$, to the right-hand-side. The reason for doing this is to bring together $q$ and $q^{-1}$ in the form of two matrices, which can be multiplied together into a single matrix.

We start with the quaternion product $qpq^{-1}$ and divide it into two parts, $qp$ and $q^{-1}$:

$$qpq^{-1} = (qp)q^{-1}.$$

The product $qp$ can be replaced by $\mathbf{L}(q)p$:

$$qpq^{-1} = \mathbf{L}(q)pq^{-1}.$$

We now have another two parts: $\mathbf{L}(q)p$ and $q^{-1}$ which can be reversed using $\mathbf{R}$ without disturbing the result:

$$qpq^{-1} = \mathbf{L}(q)pq^{-1} = \mathbf{R}(q^{-1})\mathbf{L}(q)p,$$

which has achieved our objective to move $p$ to the right-hand-side.

The next step is to compute $\mathbf{L}(q)$ and $\mathbf{R}(q^{-1})$ using $q = [s, x\mathbf{i} + y\mathbf{j} + z\mathbf{k}]$. $\mathbf{L}(q)$ is easy as it is the same as $\mathbf{L}(q_a)$:

$$\mathbf{L}(q) = \begin{bmatrix} s & -x & -y & -z \\ x & s & -z & y \\ y & z & s & -x \\ z & -y & x & s \end{bmatrix}.$$

$\mathbf{R}(q^{-1})$ is also easy, but requires converting $q_b$ in the original definition into $q^{-1}$ which is effected by reversing the signs of the vector components:

$$\mathbf{R}(q^{-1}) = \begin{bmatrix} s & x & y & z \\ -x & s & -z & y \\ -y & z & s & -x \\ -z & -y & x & s \end{bmatrix}.$$

So now we can write

$$\begin{aligned} qpq^{-1} &= \mathbf{R}(q^{-1})\mathbf{L}(q)p \\ &= \begin{bmatrix} s & x & y & z \\ -x & s & -z & y \\ -y & z & s & -x \\ -z & -y & x & s \end{bmatrix} \begin{bmatrix} s & -x & -y & -z \\ x & s & -z & y \\ y & z & s & -x \\ z & -y & x & s \end{bmatrix} \begin{bmatrix} 0 \\ x_p \\ y_p \\ z_p \end{bmatrix} \\ &= \begin{bmatrix} 1 & 0 & 0 & 0 \\ 0 & 1-2(y^2+z^2) & 2(xy-sz) & 2(xz+sy) \\ 0 & 2(xy+sz) & 1-2(x^2+z^2) & 2(yz-sx) \\ 0 & 2(xz-sy) & 2(yz+sx) & 1-2(x^2+y^2) \end{bmatrix} \begin{bmatrix} 0 \\ x_p \\ y_p \\ z_p \end{bmatrix}. \end{aligned}$$

If we ignore the first row and column, the matrix computes $\mathbf{p}'$:

$$\mathbf{p}' = \begin{bmatrix} 1-2(y^2+z^2) & 2(xy-sz) & 2(xz+sy) \\ 2(xy+sz) & 1-2(x^2+z^2) & 2(yz-sx) \\ 2(xz-sy) & 2(yz+sx) & 1-2(x^2+y^2) \end{bmatrix} \begin{bmatrix} x_p \\ y_p \\ z_p \end{bmatrix}$$

which is identical to (7.3)!

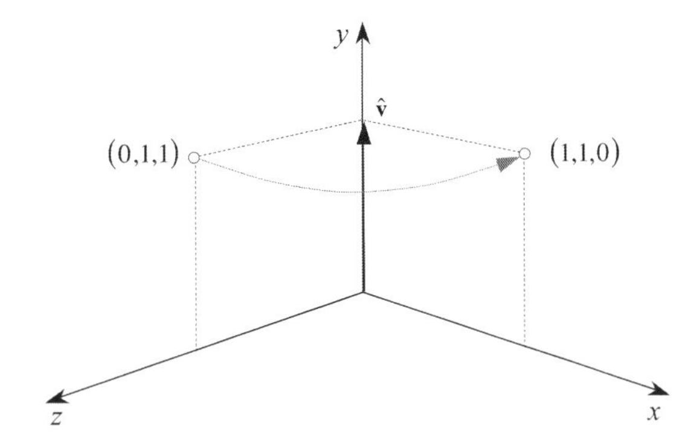


**Fig. 7.4** The point $P(0,1,1)$ is rotated 90° to $P'(1,1,0)$ about the $y$ axis

## 7.9.3 Geometric Verification

Let's illustrate the action of (7.2) by rotating the point (0, 1, 1), 90° about the $y$ axis, as shown in Fig. 7.4. The quaternion takes the form

$$q = \left[\cos\frac{\theta}{2}, \sin\frac{\theta}{2}\hat{\mathbf{v}}\right],$$

which means that $\theta = 90°$ and $\hat{\mathbf{v}} = \mathbf{j}$, therefore,

$$q = \left[\cos 45°, \sin 45°\hat{\mathbf{j}}\right].$$

Consequently,

$$s = \frac{\sqrt{2}}{2}, \qquad x = 0, \qquad y = \frac{\sqrt{2}}{2}, \qquad z = 0.$$

Substituting these values in (7.2) gives

$$\mathbf{p}' = \begin{bmatrix} 2(s^2+x^2)-1 & 2(xy-sz) & 2(xz+sy) \\ 2(xy+sz) & 2(s^2+y^2)-1 & 2(yz-sx) \\ 2(xz-sy) & 2(yz+sx) & 2(s^2+z^2)-1 \end{bmatrix} \begin{bmatrix} x_p \\ y_p \\ z_p \end{bmatrix}$$

$$\begin{bmatrix} 1 \\ 1 \\ 0 \end{bmatrix} = \begin{bmatrix} 0 & 0 & 1 \\ 0 & 1 & 0 \\ -1 & 0 & 0 \end{bmatrix} \begin{bmatrix} 0 \\ 1 \\ 1 \end{bmatrix}$$

where (0, 1, 1) is rotated to (1, 1, 0), which is correct.

So now we have a transform that rotates a point about an arbitrary axis intersecting the origin without the problems of gimbal lock associated with Euler transforms.

Before moving on, let's evaluate one more example. Let's perform a 180° rotation about a vector $\mathbf{v} = \mathbf{i} + \mathbf{k}$ passing through the origin. To begin with, we will deliberately forget to convert the vector into a unit vector, just to see what happens to the final matrix. The quaternion takes the form

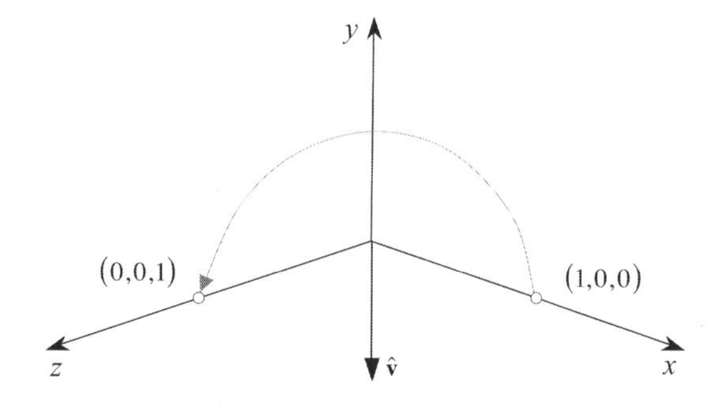


**Fig. 7.5** The point $(1,0,0)$ is rotated 180° about the vector $\hat{\mathbf{v}}$ to $(0,0,1)$

$$q = \left[\cos\frac{\theta}{2}, \sin\frac{\theta}{2}\hat{\mathbf{v}}\right],$$

but we will use $\mathbf{v}$ as specified. Therefore, with $\theta = 180°$

$$s = 0, \qquad x = 1, \qquad y = 0, \qquad z = 1.$$

Substituting these values in (7.2) gives

$$\mathbf{p}' = \begin{bmatrix} 2(s^2+x^2)-1 & 2(xy-sz) & 2(xz+sy) \\ 2(xy+sz) & 2(s^2+y^2)-1 & 2(yz-sx) \\ 2(xz-sy) & 2(yz+sx) & 2(s^2+z^2)-1 \end{bmatrix} \begin{bmatrix} x_p \\ y_p \\ z_p \end{bmatrix}$$
$$= \begin{bmatrix} 1 & 0 & 2 \\ 0 & -1 & 0 \\ 2 & 0 & 1 \end{bmatrix} \begin{bmatrix} 1 \\ 0 \\ 0 \end{bmatrix}$$

which looks nothing like a rotation matrix, and reminds us how important it is to have a unit vector to represent the axis. Let's repeat these calculations normalising the vector to $\hat{\mathbf{v}} = \frac{1}{\sqrt{2}}\mathbf{i} + \frac{1}{\sqrt{2}}\mathbf{k}$:

$$s = 0, \qquad x = \frac{1}{\sqrt{2}}, \qquad y = 0, \qquad z = \frac{1}{\sqrt{2}}.$$

Substituting these values in (7.2) gives

$$\mathbf{p}' = \begin{bmatrix} 2(s^2+x^2)-1 & 2(xy-sz) & 2(xz+sy) \\ 2(xy+sz) & 2(s^2+y^2)-1 & 2(yz-sx) \\ 2(xz-sy) & 2(yz+sx) & 2(s^2+z^2)-1 \end{bmatrix} \begin{bmatrix} x_p \\ y_p \\ z_p \end{bmatrix}$$
$$\begin{bmatrix} 0 \\ 0 \\ 1 \end{bmatrix} = \begin{bmatrix} 0 & 0 & 1 \\ 0 & -1 & 0 \\ 1 & 0 & 0 \end{bmatrix} \begin{bmatrix} 1 \\ 0 \\ 0 \end{bmatrix}$$

which not only looks like a rotation matrix, but has a determinant of 1 and rotates the point $(1,0,0)$ to $(0,0,1)$ as shown in Fig. 7.5.

## 7.10 Multiple Rotations

Say a vector or frame of reference is subjected to two rotations specified by $q_1$ followed by $q_2$. There is a temptation to convert both quaternions to their respective matrix and multiply the matrices together. However, this not the most efficient way of combining the rotations. It is best to accumulate the rotations as quaternions and then convert to matrix notation, if required. To illustrate this, consider the pure quaternion $p$ subjected to the first quaternion $q_1$:

$$q_1 p q_1^{-1},$$

followed by a second quaternion $q_2$

$$q_2\left(q_1 p q_1^{-1}\right)q_2^{-1},$$

which can be expressed as

$$(q_2 q_1)p(q_2 q_1)^{-1}.$$

Extra quaternions can be added accordingly. Let's illustrate this with two examples.

To keep things simple, the first quaternion $q_1$ rotates 30° about the $y$ axis:

$$q_1 = \left[\cos 15^\circ, \sin 15^\circ \mathbf{j}\right].$$

The second quaternion $q_2$ rotates 60° also about the $y$ axis:

$$q_2 = \left[\cos 30^\circ, \sin 30^\circ \mathbf{j}\right].$$

Together, the two quaternions rotate 90° about the $y$ axis. To accumulate these rotations, we multiply them together:

$$\begin{aligned}
q_1 q_2 &= \left[\cos 15^\circ, \sin 15^\circ \mathbf{j}\right]\left[\cos 30^\circ, \sin 30^\circ \mathbf{j}\right] \\
&= \left[\cos 15^\circ \cos 30^\circ - \sin 15^\circ \sin 30^\circ, \cos 15^\circ \sin 30^\circ \mathbf{j} + \cos 30^\circ \sin 15^\circ \mathbf{j}\right] \\
&= \left[\frac{\sqrt{2}}{2}, \frac{\sqrt{2}}{2}\mathbf{j}\right]
\end{aligned}$$

which is a quaternion that rotates 90° about the $y$ axis. Using the matrix (7.2) we have

$$\begin{aligned}
\mathbf{p}' &= \begin{bmatrix} 2(s^2+x^2)-1 & 2(xy-sz) & 2(xz+sy) \\ 2(xy+sz) & 2(s^2+y^2)-1 & 2(yz-sx) \\ 2(xz-sy) & 2(yz+sx) & 2(s^2+z^2)-1 \end{bmatrix} \begin{bmatrix} x_p \\ y_p \\ z_p \end{bmatrix} \\
&= \begin{bmatrix} 0 & 0 & 1 \\ 0 & 1 & 0 \\ -1 & 0 & 0 \end{bmatrix} \begin{bmatrix} x_p \\ y_p \\ z_p \end{bmatrix}
\end{aligned}$$

which rotates points about the $y$ axis by 90°.

For a second example, let's just evaluate the quaternions. The first quaternion $q_1$ rotates 90° about the $x$ axis, and $q_2$ rotates 90° about the $y$ axis:

$$
\begin{aligned}
q_1 &= \left[\frac{\sqrt{2}}{2}, \frac{\sqrt{2}}{2}\mathbf{i}\right] \\
q_2 &= \left[\frac{\sqrt{2}}{2}, \frac{\sqrt{2}}{2}\mathbf{j}\right] \\
p &= [0, \mathbf{i}+\mathbf{j}]
\end{aligned}
$$

therefore,

$$
\begin{aligned}
q_2 q_1 &= \left[\frac{\sqrt{2}}{2}, \frac{\sqrt{2}}{2}\mathbf{j}\right]\left[\frac{\sqrt{2}}{2}, \frac{\sqrt{2}}{2}\mathbf{i}\right] \\
&= \left[\frac{1}{2}, \frac{\sqrt{2}}{2}\frac{\sqrt{2}}{2}\mathbf{i} + \frac{\sqrt{2}}{2}\frac{\sqrt{2}}{2}\mathbf{j} - \frac{1}{2}\mathbf{k}\right] \\
&= \left[\frac{1}{2}, \frac{1}{2}\mathbf{i} + \frac{1}{2}\mathbf{j} - \frac{1}{2}\mathbf{k}\right] \\
(q_2 q_1)^{-1} &= \left[\frac{1}{2}, -\frac{1}{2}\mathbf{i} - \frac{1}{2}\mathbf{j} + \frac{1}{2}\mathbf{k}\right] \\
(q_2 q_1)p &= \left[\frac{1}{2}, \frac{1}{2}\mathbf{i} + \frac{1}{2}\mathbf{j} - \frac{1}{2}\mathbf{k}\right][0, \mathbf{i}+\mathbf{j}] \\
&= \left[-\frac{1}{2} - \frac{1}{2}, \frac{1}{2}(\mathbf{i}+\mathbf{j}) + \frac{1}{2}\mathbf{i} - \frac{1}{2}\mathbf{j}\right] \\
&= [-1, \mathbf{i}] \\
(q_2 q_1)p(q_2 q_1)^{-1} &= [-1, \mathbf{i}]\left[\frac{1}{2}, -\frac{1}{2}\mathbf{i} - \frac{1}{2}\mathbf{j} + \frac{1}{2}\mathbf{k}\right] \\
&= \left[-\frac{1}{2} + \frac{1}{2}, \frac{1}{2}\mathbf{i} + \frac{1}{2}\mathbf{j} - \frac{1}{2}\mathbf{k} + \frac{1}{2}\mathbf{i} - \frac{1}{2}\mathbf{j} - \frac{1}{2}\mathbf{k}\right] \\
&= [0, \mathbf{i} - \mathbf{k}].
\end{aligned}
$$

Thus the point $(1, 1, 0)$ is rotated to $(1, 0, -1)$, which is correct.

## 7.11 Eigenvalue and Eigenvector

Although there is no doubt that (7.2) is a rotation matrix, we can secure further evidence by calculating its eigenvalue and eigenvector. The eigenvalue should be $\theta$

where

$$\mathrm{Tr}(qpq^{-1}) = 1 + 2\cos\theta.$$

and Tr is the trace function, which is the sum of the diagonal elements of a matrix.

The trace of (7.2) is

$$\begin{aligned}
\mathrm{Tr}(qpq^{-1}) &= 2(s^2 + x^2) - 1 + 2(s^2 + y^2) - 1 + 2(s^2 + z^2) - 1 \\
&= 4s^2 + 2(s^2 + x^2 + y^2 + z^2) - 3 \\
&= 4s^2 - 1 \\
&= 4\cos^2\frac{1}{2}\theta - 1 \\
&= 4\cos\theta + 4\sin^2\frac{1}{2}\theta - 1 \\
&= 4\cos\theta + 2 - 2\cos\theta - 1 \\
&= 1 + 2\cos\theta
\end{aligned}$$

and

$$\cos\theta = \frac{1}{2}\left(\mathrm{Tr}(qpq^{-1}) - 1\right).$$

To compute the eigenvector of (7.2) we use the three equations derived in Chap. 6:

$$\begin{aligned}
x_v &= (a_{22} - 1)(a_{33} - 1) - a_{23}a_{32} \\
y_v &= (a_{33} - 1)(a_{11} - 1) - a_{31}a_{13} \\
z_v &= (a_{11} - 1)(a_{22} - 1) - a_{12}a_{21}.
\end{aligned}$$

Therefore,

$$\begin{aligned}
x_v &= \left(2(s^2 + y^2) - 2\right)\left(2(s^2 + z^2) - 2\right) - 2(yz - sx)2(yz + sx) \\
&= 4(s^2 + y^2 - 1)(s^2 + z^2 - 1) - 4(y^2z^2 - s^2x^2) \\
&= 4\left((x^2 + z^2)(x^2 + y^2) - y^2z^2 + s^2x^2\right) \\
&= 4(x^4 + x^2y^2 + x^2z^2 + z^2y^2 - y^2z^2 + s^2x^2) \\
&= 4x^2(s^2 + x^2 + y^2 + z^2) \\
&= 4x^2.
\end{aligned}$$

Similarly, $y_v = 4y^2$ and $z_v = 4z^2$, which confirm that the eigenvector has components associated with the quaternion's vector. The square terms should be no surprise, as the triple $qpq^{-1}$ includes the product of three quaternions.

Let's test these formulae with the matrix associated with Fig. 7.5, which rotates a point 180° about the vector $\hat{\mathbf{v}} = \frac{1}{\sqrt{2}}\mathbf{i} + \frac{1}{\sqrt{2}}\mathbf{k}$:

$$\mathbf{M} = \begin{bmatrix} a_{11} & a_{12} & a_{13} \\ a_{21} & a_{22} & a_{23} \\ a_{31} & a_{32} & a_{33} \end{bmatrix} = \begin{bmatrix} 0 & 0 & 1 \\ 0 & -1 & 0 \\ 1 & 0 & 0 \end{bmatrix},$$

therefore,

$$\begin{aligned} x_v &= -2 \times -1 - 0 = 2 \\ y_v &= -1 \times -1 - 1 \times 1 = 0 \\ z_v &= -1 \times -2 - 0 = 2 \end{aligned}$$

which confirms that the eigenvector is $2\mathbf{i} + 2\mathbf{k}$.

Next, $\operatorname{Tr}(\mathbf{M}) = -1$, therefore

$$\begin{aligned} \cos\theta &= \frac{1}{2}\left(\operatorname{Tr}\left(qpq^{-1}\right) - 1\right) \\ &= \frac{1}{2}\left((-1) - 1\right) \\ &= -1 \\ \theta &= \pm 180^\circ \end{aligned}$$

which agrees with the previous results.

## 7.12 Rotating About an Off-Set Axis

Now that we have a matrix to represent a quaternion rotor, we can employ it to resolve problems such as rotating a point about an off-set axis using the same techniques associated with normal rotation transforms. For example, in Chap. 6 we used the following notation

$$\begin{bmatrix} x' \\ y' \\ z' \\ 1 \end{bmatrix} = [\mathbf{T}_{t_x,0,t_z}][\mathbf{R}_{\beta,y}][\mathbf{T}_{-t_x,0,-t_z}] \begin{bmatrix} x \\ y \\ z \\ 1 \end{bmatrix},$$

to rotate a point about a fixed axis parallel with the $y$ axis. Therefore, by substituting the matrix $qpq^{-1}$ for $\mathbf{R}_{\beta,y}$ we have

$$\begin{bmatrix} x' \\ y' \\ z' \\ 1 \end{bmatrix} = [\mathbf{T}_{t_x,0,t_z}]\left(qpq^{-1}\right)[\mathbf{T}_{-t_x,0,-t_z}] \begin{bmatrix} x \\ y \\ z \\ 1 \end{bmatrix}.$$

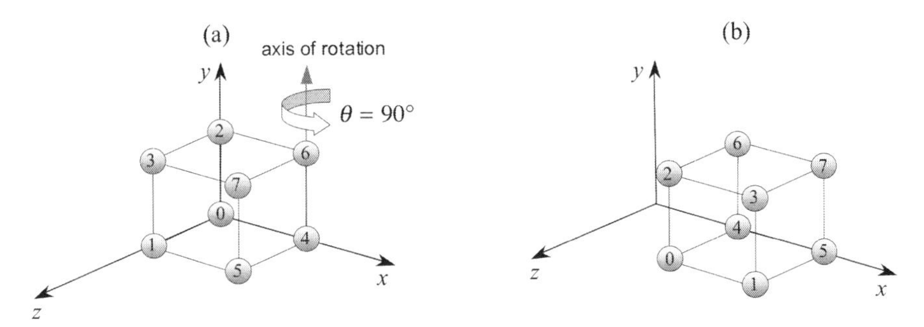


**Fig. 7.6** The cube is rotated 90° about the axis intersecting vertices 4 and 6

Let's test this by rotating our unit cube 90° about the vertical axis intersecting vertices 4 and 6 as shown in Fig. 7.6. The unit-norm quaternion to achieve this is

$$q = \left[\cos 45°, \sin 45°\mathbf{j}\right],$$

with the pure quaternion

$$p = [0, \mathbf{p}].$$

Consequently,

$$s = \frac{\sqrt{2}}{2}, \qquad x = 0, \qquad y = \frac{\sqrt{2}}{2}, \qquad z = 0$$

and using (7.2) in a homogeneous form we have

$$\mathbf{p}' = \begin{bmatrix} 2(s^2+x^2)-1 & 2(xy-sz) & 2(xz+sy) & 0 \\ 2(xy+sz) & 2(s^2+y^2)-1 & 2(yz-sx) & 0 \\ 2(xz-sy) & 2(yz+sx) & 2(s^2+z^2)-1 & 0 \\ 0 & 0 & 0 & 1 \end{bmatrix} \begin{bmatrix} x_p \\ y_p \\ z_p \\ 1 \end{bmatrix}$$

$$= \begin{bmatrix} 0 & 0 & 1 & 0 \\ 0 & 1 & 0 & 0 \\ -1 & 0 & 0 & 0 \\ 0 & 0 & 0 & 1 \end{bmatrix} \begin{bmatrix} x_p \\ y_p \\ z_p \\ 1 \end{bmatrix}.$$

The other two matrices are

$$\mathbf{T}_{-t_x,0,0} = \begin{bmatrix} 1 & 0 & 0 & -1 \\ 0 & 1 & 0 & 0 \\ 0 & 0 & 1 & 0 \\ 0 & 0 & 0 & 1 \end{bmatrix}$$

$$\mathbf{T}_{t_x,0,0} = \begin{bmatrix} 1 & 0 & 0 & 1 \\ 0 & 1 & 0 & 0 \\ 0 & 0 & 1 & 0 \\ 0 & 0 & 0 & 1 \end{bmatrix}.$$

Multiplying these three matrices together creates

$$\begin{bmatrix} 0 & 0 & 1 & 1 \\ 0 & 1 & 0 & 0 \\ -1 & 0 & 0 & 1 \\ 0 & 0 & 0 & 1 \end{bmatrix}. \tag{7.6}$$

Although not mathematically correct, the following statement shows the matrix (7.6) and the array of coordinates representing a unit cube, followed by the rotated cube's coordinates.

$$\begin{bmatrix} 0 & 0 & 1 & 1 \\ 0 & 1 & 0 & 0 \\ -1 & 0 & 0 & 1 \\ 0 & 0 & 0 & 1 \end{bmatrix} \begin{bmatrix} 0 & 0 & 0 & 0 & 1 & 1 & 1 & 1 \\ 0 & 0 & 1 & 1 & 0 & 0 & 1 & 1 \\ 0 & 1 & 0 & 1 & 0 & 1 & 0 & 1 \\ 1 & 1 & 1 & 1 & 1 & 1 & 1 & 1 \end{bmatrix}$$
$$= \begin{bmatrix} 1 & 2 & 1 & 2 & 1 & 2 & 1 & 2 \\ 0 & 0 & 1 & 1 & 0 & 0 & 1 & 1 \\ 1 & 1 & 1 & 1 & 0 & 0 & 0 & 0 \\ 1 & 1 & 1 & 1 & 1 & 1 & 1 & 1 \end{bmatrix}.$$

These coordinates are confirmed by Fig. 7.6.

## 7.13 Frames of Reference

The product $qpq^{-1}$ is used for rotating points about the vector associated with the quaternion $q$, whereas the triple $q^{-1}pq$ can be used for rotating points about the same vector in the opposite direction. But this reverse rotation is also equivalent to a change of frame of reference. To demonstrate this, consider the problem of rotating the frame of reference $180°$ about $\mathbf{i}+\mathbf{k}$ as shown in Fig. 7.7. The unit-norm quaternion for such a rotation is

$$\begin{aligned} q &= \left[\cos 90°, \sin 90°\left(\frac{1}{\sqrt{2}}\mathbf{i} + \frac{1}{\sqrt{2}}\mathbf{k}\right)\right] \\ &= \left[0, \frac{\sqrt{2}}{2}\mathbf{i} + \frac{\sqrt{2}}{2}\mathbf{k}\right]. \end{aligned}$$

Consequently,

$$s = 0, \qquad x = \frac{\sqrt{2}}{2}, \qquad y = 0, \qquad z = \frac{\sqrt{2}}{2}.$$

Substituting these values in (7.4) we obtain

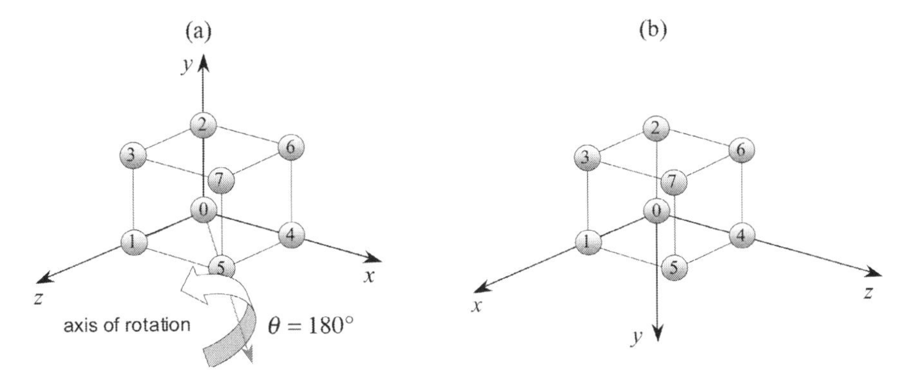


**Fig. 7.7** The frame is rotated 180° about the vector $\mathbf{i}+\mathbf{k}$

$$
\begin{aligned}
q^{-1}pq &= \begin{bmatrix} 2(s^2+x^2)-1 & 2(xy+sz) & 2(xz-sy) \\ 2(xy-sz) & 2(s^2+y^2)-1 & 2(yz+sx) \\ 2(xz+sy) & 2(yz-sx) & 2(s^2+z^2)-1 \end{bmatrix} \begin{bmatrix} x_p \\ y_p \\ z_p \end{bmatrix} \\
&= \begin{bmatrix} 0 & 0 & 1 \\ 0 & -1 & 0 \\ 1 & 0 & 0 \end{bmatrix} \begin{bmatrix} x_p \\ y_p \\ z_p \end{bmatrix}
\end{aligned}
$$

which, if used to process the coordinates of our unit cube, produces

$$
\begin{aligned}
&\begin{bmatrix} 0 & 0 & 1 \\ 0 & -1 & 0 \\ 1 & 0 & 0 \end{bmatrix} \begin{bmatrix} 0 & 0 & 0 & 0 & 1 & 1 & 1 & 1 \\ 0 & 0 & 1 & 1 & 0 & 0 & 1 & 1 \\ 0 & 1 & 0 & 1 & 0 & 1 & 0 & 1 \end{bmatrix} \\
&= \begin{bmatrix} 0 & 1 & 0 & 1 & 0 & 1 & 0 & 1 \\ 0 & 0 & -1 & -1 & 0 & 0 & -1 & -1 \\ 0 & 0 & 0 & 0 & 1 & 1 & 1 & 1 \end{bmatrix}.
\end{aligned}
$$

This scenario is shown in Fig. 7.7.

## 7.14 Euler Angles to Quaternion

In Chap. 6 we discovered that the rotation transforms $\mathbf{R}_{\alpha,x}$, $\mathbf{R}_{\beta,y}$ and $\mathbf{R}_{\gamma,z}$ can be combined to create twelve, triple combinations to represent a composite rotation. Now let's see how such a transform is represented by a quaternion.

To demonstrate the technique we must choose one of the twelve combinations, then the same technique can be used to convert other combinations. For example, let's choose the sequence $\mathbf{R}_{\gamma,z}\mathbf{R}_{\beta,y}\mathbf{R}_{\alpha,x}$ where the equivalent quaternions are

$$
q_x = \left[\cos\frac{\alpha}{2}, \sin\frac{\alpha}{2}\mathbf{i}\right]
$$

$$q_y = \left[\cos\frac{\beta}{2}, \sin\frac{\beta}{2}\mathbf{j}\right]$$

$$q_z = \left[\cos\frac{\gamma}{2}, \sin\frac{\gamma}{2}\mathbf{k}\right]$$

and

$$q = q_z q_y q_x. \tag{7.7}$$

Expanding (7.7):

$$\begin{aligned}
q &= \left[\cos\frac{\gamma}{2}, \sin\frac{\gamma}{2}\mathbf{k}\right]\left[\cos\frac{\beta}{2}, \sin\frac{\beta}{2}\mathbf{j}\right]\left[\cos\frac{\alpha}{2}, \sin\frac{\alpha}{2}\mathbf{i}\right] \\
&= \left[\cos\frac{\gamma}{2}\cos\frac{\beta}{2},\right. \\
&\quad \left.\cos\frac{\gamma}{2}\sin\frac{\beta}{2}\mathbf{j} + \cos\frac{\beta}{2}\sin\frac{\gamma}{2}\mathbf{k} - \sin\frac{\gamma}{2}\sin\frac{\beta}{2}\mathbf{i}\right]\left[\cos\frac{\alpha}{2}, \sin\frac{\alpha}{2}\mathbf{i}\right] \\
&= \left[\cos\frac{\gamma}{2}\cos\frac{\beta}{2}\cos\frac{\alpha}{2} + \sin\frac{\gamma}{2}\sin\frac{\beta}{2}\sin\frac{\alpha}{2},\right. \\
&\quad \cos\frac{\gamma}{2}\cos\frac{\beta}{2}\sin\frac{\alpha}{2}\mathbf{i} + \cos\frac{\alpha}{2}\cos\frac{\gamma}{2}\sin\frac{\beta}{2}\mathbf{j} + \cos\frac{\alpha}{2}\cos\frac{\beta}{2}\sin\frac{\gamma}{2}\mathbf{k} \\
&\quad \left.- \cos\frac{\alpha}{2}\sin\frac{\gamma}{2}\sin\frac{\beta}{2}\mathbf{i} - \cos\frac{\gamma}{2}\sin\frac{\beta}{2}\sin\frac{\alpha}{2}\mathbf{k} + \cos\frac{\beta}{2}\sin\frac{\gamma}{2}\sin\frac{\alpha}{2}\mathbf{j}\right] \\
&= \left[\cos\frac{\gamma}{2}\cos\frac{\beta}{2}\cos\frac{\alpha}{2} + \sin\frac{\gamma}{2}\sin\frac{\beta}{2}\sin\frac{\alpha}{2},\right. \\
&\quad \left(\cos\frac{\gamma}{2}\cos\frac{\beta}{2}\sin\frac{\alpha}{2} - \cos\frac{\alpha}{2}\sin\frac{\gamma}{2}\sin\frac{\beta}{2}\right)\mathbf{i} \\
&\quad + \left(\cos\frac{\alpha}{2}\cos\frac{\gamma}{2}\sin\frac{\beta}{2} + \cos\frac{\beta}{2}\sin\frac{\gamma}{2}\sin\frac{\alpha}{2}\right)\mathbf{j} \\
&\quad \left.+ \left(\cos\frac{\alpha}{2}\cos\frac{\beta}{2}\sin\frac{\gamma}{2} - \cos\frac{\gamma}{2}\sin\frac{\beta}{2}\sin\frac{\alpha}{2}\right)\mathbf{k}\right].
\end{aligned}$$

Now let's place the angles in a consistent sequence:

$$s = \cos\frac{\gamma}{2}\cos\frac{\beta}{2}\cos\frac{\alpha}{2} + \sin\frac{\gamma}{2}\sin\frac{\beta}{2}\sin\frac{\alpha}{2}$$

$$x_q = \cos\frac{\gamma}{2}\cos\frac{\beta}{2}\sin\frac{\alpha}{2} - \sin\frac{\gamma}{2}\sin\frac{\beta}{2}\cos\frac{\alpha}{2}$$

$$y_q = \cos\frac{\gamma}{2}\sin\frac{\beta}{2}\cos\frac{\alpha}{2} + \sin\frac{\gamma}{2}\cos\frac{\beta}{2}\sin\frac{\alpha}{2}$$

$$z_q = \sin\frac{\gamma}{2}\cos\frac{\beta}{2}\cos\frac{\alpha}{2} - \cos\frac{\gamma}{2}\sin\frac{\beta}{2}\sin\frac{\alpha}{2}$$

where

$$q = [s, x_q\mathbf{i} + y_q\mathbf{j} + z_q\mathbf{k}]. \tag{7.8}$$

Let's test (7.8). We start with the three rotation transforms

$$\mathbf{R}_{\alpha,x} = \begin{bmatrix} 1 & 0 & 0 \\ 0 & \cos\alpha & -\sin\alpha \\ 0 & \sin\alpha & \cos\alpha \end{bmatrix}$$

$$\mathbf{R}_{\beta,y} = \begin{bmatrix} \cos\beta & 0 & \sin\beta \\ 0 & 1 & 0 \\ -\sin\beta & 0 & \cos\beta \end{bmatrix}$$

$$\mathbf{R}_{\gamma,z} = \begin{bmatrix} \cos\gamma & -\sin\gamma & 0 \\ \sin\gamma & \cos\gamma & 0 \\ 0 & 0 & 1 \end{bmatrix}.$$

Then

$$\mathbf{R}_{\gamma,z}\mathbf{R}_{\beta,y}\mathbf{R}_{\alpha,x}$$
$$= \begin{bmatrix} \cos\gamma\cos\beta & -\sin\gamma\cos\alpha + \cos\gamma\sin\beta\sin\alpha & \sin\gamma\sin\alpha + \cos\gamma\sin\beta\cos\alpha \\ \sin\gamma\cos\beta & \cos\gamma\cos\alpha + \sin\gamma\sin\beta\sin\alpha & -\cos\gamma\sin\alpha + \sin\gamma\sin\beta\cos\alpha \\ -\sin\beta & \cos\beta\sin\alpha & \cos\beta\cos\alpha \end{bmatrix}.$$

Let's make $\alpha = \beta = \gamma = 90°$, then

$$\mathbf{R}_{90°,z}\mathbf{R}_{90°,y}\mathbf{R}_{90°,x} = \begin{bmatrix} 0 & 0 & 1 \\ 0 & 1 & 0 \\ -1 & 0 & 0 \end{bmatrix},$$

which rotates points 90° about the $y$ axis:

$$\begin{bmatrix} 1 \\ 1 \\ 0 \end{bmatrix} = \begin{bmatrix} 0 & 0 & 1 \\ 0 & 1 & 0 \\ -1 & 0 & 0 \end{bmatrix} \begin{bmatrix} 0 \\ 1 \\ 1 \end{bmatrix}.$$

Now let's evaluate (7.8):

$$\begin{aligned} s &= \cos\frac{\gamma}{2}\cos\frac{\beta}{2}\cos\frac{\alpha}{2} + \sin\frac{\gamma}{2}\sin\frac{\beta}{2}\sin\frac{\alpha}{2} \\ &= \frac{\sqrt{2}}{2}\frac{\sqrt{2}}{2}\frac{\sqrt{2}}{2} + \frac{\sqrt{2}}{2}\frac{\sqrt{2}}{2}\frac{\sqrt{2}}{2} \\ &= \frac{\sqrt{2}}{2} \end{aligned}$$

$$\begin{aligned}
x_q &= \cos\frac{\gamma}{2}\cos\frac{\beta}{2}\sin\frac{\alpha}{2} - \sin\frac{\gamma}{2}\sin\frac{\beta}{2}\cos\frac{\alpha}{2} \\
&= 0 \\
y_q &= \cos\frac{\gamma}{2}\sin\frac{\beta}{2}\cos\frac{\alpha}{2} + \sin\frac{\gamma}{2}\cos\frac{\beta}{2}\sin\frac{\alpha}{2} \\
&= \frac{\sqrt{2}}{2}\frac{\sqrt{2}}{2}\frac{\sqrt{2}}{2} + \frac{\sqrt{2}}{2}\frac{\sqrt{2}}{2}\frac{\sqrt{2}}{2} \\
&= \frac{\sqrt{2}}{2} \\
z_q &= \sin\frac{\gamma}{2}\cos\frac{\beta}{2}\cos\frac{\alpha}{2} - \cos\frac{\gamma}{2}\sin\frac{\beta}{2}\sin\frac{\alpha}{2} \\
&= 0
\end{aligned}$$

and

$$q = \left[\frac{\sqrt{2}}{2}, \frac{\sqrt{2}}{2}\mathbf{j}\right],$$

which is a quaternion that also rotates points 90° about the $y$ axis.

## 7.15 Worked Examples

Here are some further worked examples that employ the ideas described above.

*Example 1* Use $qp$ to rotate $p = [0, \mathbf{j}]$ 90° about the $x$ axis.

For this to work $q$ must be orthogonal to $p$:

$$\begin{aligned}
q &= [\cos\theta, \sin\theta\mathbf{i}] \\
&= [0, \mathbf{i}]
\end{aligned}$$

and

$$\begin{aligned}
p' &= qp \\
&= [0, \mathbf{i}][0, \mathbf{j}] \\
&= [0, \mathbf{k}].
\end{aligned}$$

*Example 2* Use $qpq^{-1}$ to rotate $p = [0, \mathbf{j}]$ 90° about the $x$ axis.

For this to work:

$$q = \left[\cos\frac{\theta}{2}, \sin\frac{\theta}{2}\mathbf{i}\right]$$

$$= \left[\frac{\sqrt{2}}{2}, \frac{\sqrt{2}}{2}\mathbf{i}\right]$$

and

$$\begin{aligned}
p' &= qpq^{-1} \\
&= \left[\frac{\sqrt{2}}{2}, \frac{\sqrt{2}}{2}\mathbf{i}\right][0, \mathbf{j}]\left[\frac{\sqrt{2}}{2}, -\frac{\sqrt{2}}{2}\mathbf{i}\right] \\
&= \left[0, \frac{\sqrt{2}}{2}\mathbf{j} + \frac{\sqrt{2}}{2}\mathbf{k}\right]\left[\frac{\sqrt{2}}{2}, -\frac{\sqrt{2}}{2}\mathbf{i}\right] \\
&= \left[0, \frac{\sqrt{2}}{2}\left(\frac{\sqrt{2}}{2}\mathbf{j} + \frac{\sqrt{2}}{2}\mathbf{k}\right) + \frac{1}{2}\mathbf{j} + \frac{1}{2}\mathbf{k}\right] \\
&= \left[0, \frac{1}{2}\mathbf{j} + \frac{1}{2}\mathbf{k} - \frac{1}{2}\mathbf{j} + \frac{1}{2}\mathbf{k}\right] \\
&= [0, \mathbf{k}]
\end{aligned}$$

which agrees with the answer for Example 1.

*Example 3* Evaluate the triple $qpq^{-1}$ for $p = [0, \mathbf{p}]$ and $q = [\cos\frac{\theta}{2}, \sin\frac{\theta}{2}\mathbf{v}]$, where $\theta = 360°$.

$$\begin{aligned}
q &= [-1, \mathbf{0}] \\
qpq^{-1} &= [-1, \mathbf{0}][0, \mathbf{p}][-1, \mathbf{0}] \\
&= [0, -\mathbf{p}][-1, \mathbf{0}] \\
&= [0, \mathbf{p}]
\end{aligned}$$

which confirms that the vector remains unmoved, as expected.

*Example 4* Compute the matrix (7.2) for $q = [\frac{1}{2}, \frac{\sqrt{3}}{2}\mathbf{k}]$, and find its eigenvector and eigenvalue.

From $q$:

$$s = \frac{1}{2}, \qquad x = 0, \qquad y = 0, \qquad z = \frac{\sqrt{3}}{2}$$

$$\begin{aligned}
\mathbf{p}' &= \begin{bmatrix} 2(s^2+x^2)-1 & 2(xy-sz) & 2(xz+sy) \\ 2(xy+sz) & 2(s^2+y^2)-1 & 2(yz-sx) \\ 2(xz-sy) & 2(yz+sx) & 2(s^2+z^2)-1 \end{bmatrix} \begin{bmatrix} x_p \\ y_p \\ z_p \end{bmatrix} \\
&= \begin{bmatrix} -\frac{1}{2} & -\frac{\sqrt{3}}{2} & 0 \\ \frac{\sqrt{3}}{2} & -\frac{1}{2} & 0 \\ 0 & 0 & 1 \end{bmatrix} \begin{bmatrix} x_p \\ y_p \\ z_p \end{bmatrix}.
\end{aligned}$$

If we plug in the point $(1, 0, 0)$ it is rotated about the $z$ axis by $120°$:

$$\begin{bmatrix} -\frac{1}{2} \\ \frac{\sqrt{3}}{2} \\ 1 \end{bmatrix} = \begin{bmatrix} -\frac{1}{2} & -\frac{\sqrt{3}}{2} & 0 \\ \frac{\sqrt{3}}{2} & -\frac{1}{2} & 0 \\ 0 & 0 & 1 \end{bmatrix} \begin{bmatrix} 1 \\ 0 \\ 0 \end{bmatrix}.$$

Using

$$\begin{aligned} \cos\theta &= \frac{1}{2}\big(\mathrm{Tr}\big(qpq^{-1}\big) - 1\big) \\ &= \frac{1}{2}(0 - 1) \\ \theta &= 120°. \end{aligned}$$

Using

$$\begin{aligned} x_v &= (a_{22} - 1)(a_{33} - 1) - a_{23}a_{32} \\ &= \left(-\frac{3}{2}\right)(0) - 0 \\ &= 0 \\ y_v &= (a_{33} - 1)(a_{11} - 1) - a_{31}a_{13} \\ &= (0)\left(-\frac{3}{2}\right) - 0 \\ &= 0 \\ z_v &= (a_{11} - 1)(a_{22} - 1) - a_{12}a_{21} \\ &= \left(-\frac{3}{2}\right)\left(-\frac{3}{2}\right) + \frac{\sqrt{3}}{2}\frac{\sqrt{3}}{2} \\ &= 3 \end{aligned}$$

which makes the eigenvector $3\mathbf{k}$ and the eigenvalue $120°$.

*Example 5* Convert the given matrix into a quaternion and identify its function.

$$\mathbf{M} = \begin{bmatrix} 0 & 0 & 1 \\ 0 & 1 & 0 \\ -1 & 0 & 0 \end{bmatrix},$$

therefore,

$$s = \frac{1}{2}\sqrt{1 + a_{11} + a_{22} + a_{33}}$$

$$
\begin{aligned}
&= \frac{1}{2}\sqrt{1+0+1+0} = \frac{\sqrt{2}}{2} \\
x &= \frac{1}{4s}(a_{32} - a_{23}) \\
&= \frac{\sqrt{2}}{4}(0+0) = 0 \\
y &= \frac{1}{4s}(a_{13} - a_{31}) \\
&= \frac{\sqrt{2}}{4}(1+1) = \frac{\sqrt{2}}{2} \\
z &= \frac{1}{4s}(a_{21} - a_{12}) \\
&= \frac{\sqrt{2}}{4}(0+0) = 0
\end{aligned}
$$

which is the quaternion $[\frac{\sqrt{2}}{2}, \frac{\sqrt{2}}{2}\mathbf{j}]$ and is a rotation of 90° about the $y$ axis.

## 7.16 Summary

This chapter has demonstrated how unit-norm quaternions can be used to rotate a vector about a quaternion's vector. The product $qpq^{-1}$—discovered by Hamilton and Cayley—works for all orientations between a quaternion and a vector. It is also relatively easy to compute. We also saw that the product can be represented as a matrix, which can be integrated with other matrices.

The reverse product $q^{-1}pq$ reverses the angle of rotation, and is equivalent to changing the sign of the rotation angle in $qpq^{-1}$. Consequently, it can be used to rotate a frame of reference in the same direction as $qpq^{-1}$.

# Chapter 8
# Conclusion

The objective of this book was to show the reader how groups of equations can be expressed using matrices. The notation is very compact and permits one to identify the action of the matrix transform. Matrices also provide a useful structure for storing and communicating transforms within a computer system, especially at the interface with a graphics processor.

I confined the book to the 2D and 3D matrix transforms found in computer games and animation software. However, matrix notation is widely used in computer graphics to compute perspective views, curves and surfaces, etc. Hopefully, after reading this book, the reader will understand the direct link between algebra and matrices, and appreciate the elegance matrix notation brings to the design of computer graphics algorithms.

J. Vince, *Matrix Transforms for Computer Games and Animation*,
DOI 10.1007/978-1-4471-4321-5_8, © Springer-Verlag London 2012

# Appendix
# Composite Point Rotation Sequences

## A.1 Euler Rotations

In Chap. 6 we considered composite Euler rotations comprising individual rotations about the $x$, $y$ and $z$ axes such as $\mathbf{R}_{\gamma,x}\mathbf{R}_{\beta,y}\mathbf{R}_{\alpha,z}$ and $\mathbf{R}_{\gamma,z}\mathbf{R}_{\beta,y}\mathbf{R}_{\alpha,x}$. However, there is nothing preventing us from creating other combinations such as $\mathbf{R}_{\gamma,x}\mathbf{R}_{\beta,y}\mathbf{R}_{\alpha,x}$ or $\mathbf{R}_{\gamma,z}\mathbf{R}_{\beta,y}\mathbf{R}_{\alpha,z}$ that do not include two consecutive rotations about the same axis. In all, there are twelve possible combinations:

$$\begin{array}{llll}
\mathbf{R}_{\gamma,x}\mathbf{R}_{\beta,y}\mathbf{R}_{\alpha,x} & \mathbf{R}_{\gamma,x}\mathbf{R}_{\beta,y}\mathbf{R}_{\alpha,z} & \mathbf{R}_{\gamma,x}\mathbf{R}_{\beta,z}\mathbf{R}_{\alpha,x} & \mathbf{R}_{\gamma,x}\mathbf{R}_{\beta,z}\mathbf{R}_{\alpha,y}\\
\mathbf{R}_{\gamma,y}\mathbf{R}_{\beta,x}\mathbf{R}_{\alpha,y} & \mathbf{R}_{\gamma,y}\mathbf{R}_{\beta,x}\mathbf{R}_{\alpha,z} & \mathbf{R}_{\gamma,y}\mathbf{R}_{\beta,z}\mathbf{R}_{\alpha,x} & \mathbf{R}_{\gamma,y}\mathbf{R}_{\beta,z}\mathbf{R}_{\alpha,y}\\
\mathbf{R}_{\gamma,z}\mathbf{R}_{\beta,x}\mathbf{R}_{\alpha,y} & \mathbf{R}_{\gamma,z}\mathbf{R}_{\beta,x}\mathbf{R}_{\alpha,z} & \mathbf{R}_{\gamma,z}\mathbf{R}_{\beta,y}\mathbf{R}_{\alpha,x} & \mathbf{R}_{\gamma,z}\mathbf{R}_{\beta,y}\mathbf{R}_{\alpha,z}
\end{array}$$

which we now cover in detail.

For each combination there are three Euler rotation matrices, the resulting composite matrix, a matrix where the three angles equal 90°, the coordinates of the rotated unit cube, the axis and angle of rotation and a figure illustrating the stages of rotation. To compute the axis of rotation $[v_1 \quad v_2 \quad v_3]^{\mathrm{T}}$ we use

$$\begin{aligned}
v_1 &= (a_{22} - 1)(a_{33} - 1) - a_{23}a_{32}\\
v_2 &= (a_{33} - 1)(a_{11} - 1) - a_{31}a_{13}\\
v_3 &= (a_{11} - 1)(a_{22} - 1) - a_{12}a_{21}
\end{aligned}$$

where

$$\mathbf{R} = \begin{bmatrix} a_{11} & a_{12} & a_{13}\\ a_{21} & a_{22} & a_{23}\\ a_{31} & a_{32} & a_{33} \end{bmatrix},$$

and for the angle of rotation $\delta$ we use

$$\cos\delta = \frac{1}{2}\big(\mathrm{Tr}(\mathbf{R}) - 1\big).$$

J. Vince, *Matrix Transforms for Computer Games and Animation*,
DOI 10.1007/978-1-4471-4321-5, © Springer-Verlag London 2012

We begin by defining the three principal Euler rotations:

$$\text{rotate } \alpha \text{ about the } x\text{-axis} \quad \mathbf{R}_{\alpha,x} = \begin{bmatrix} 1 & 0 & 0 \\ 0 & c_\alpha & -s_\alpha \\ 0 & s_\alpha & c_\alpha \end{bmatrix}$$

$$\text{rotate } \beta \text{ about the } y\text{-axis} \quad \mathbf{R}_{\beta,y} = \begin{bmatrix} c_\beta & 0 & s_\beta \\ 0 & 1 & 0 \\ -s_\beta & 0 & c_\beta \end{bmatrix}$$

$$\text{rotate } \gamma \text{ about the } z\text{-axis} \quad \mathbf{R}_{\gamma,z} = \begin{bmatrix} c_\gamma & -s_\gamma & 0 \\ s_\gamma & c_\gamma & 0 \\ 0 & 0 & 1 \end{bmatrix}$$

where $c_\alpha = \cos\alpha$ and $s_\alpha = \sin\alpha$, etc.

Remember that the right-most transform is applied first and the left-most transform last. In terms of angles, the sequence is always $\alpha$, $\beta$, $\gamma$.

For each composite transform you can verify that when $\alpha = \beta = \gamma = 0$ the result is the identity transform **I**.

We now examine the twelve combinations in turn.

## A.2 $\mathbf{R}_{\gamma,x}\mathbf{R}_{\beta,y}\mathbf{R}_{\alpha,x}$

$$\mathbf{R}_{\gamma,x}\mathbf{R}_{\beta,y}\mathbf{R}_{\alpha,x} = \begin{bmatrix} 1 & 0 & 0 \\ 0 & c_\gamma & -s_\gamma \\ 0 & s_\gamma & c_\gamma \end{bmatrix} \begin{bmatrix} c_\beta & 0 & s_\beta \\ 0 & 1 & 0 \\ -s_\beta & 0 & c_\beta \end{bmatrix} \begin{bmatrix} 1 & 0 & 0 \\ 0 & c_\alpha & -s_\alpha \\ 0 & s_\alpha & c_\alpha \end{bmatrix}$$

$$= \begin{bmatrix} c_\beta & s_\beta s_\alpha & s_\beta c_\alpha \\ s_\gamma s_\beta & (c_\gamma c_\alpha - s_\gamma c_\beta s_\alpha) & (-c_\gamma s_\alpha - s_\gamma c_\beta c_\alpha) \\ -c_\gamma s_\beta & (s_\gamma c_\alpha + c_\gamma c_\beta s_\alpha) & (-s_\gamma s_\alpha + c_\gamma c_\beta c_\alpha) \end{bmatrix}$$

$$\mathbf{R}_{90^\circ,x}\mathbf{R}_{90^\circ,y}\mathbf{R}_{90^\circ,x} = \begin{bmatrix} 0 & 1 & 0 \\ 1 & 0 & 0 \\ 0 & 0 & -1 \end{bmatrix}$$

$$\begin{bmatrix} 0 & 1 & 0 \\ 1 & 0 & 0 \\ 0 & 0 & -1 \end{bmatrix} \begin{bmatrix} 0 & 0 & 0 & 0 & 1 & 1 & 1 & 1 \\ 0 & 0 & 1 & 1 & 0 & 0 & 1 & 1 \\ 0 & 1 & 0 & 1 & 0 & 1 & 0 & 1 \end{bmatrix}$$

$$= \begin{bmatrix} 0 & 0 & 1 & 1 & 0 & 0 & 1 & 1 \\ 0 & 0 & 0 & 0 & 1 & 1 & 1 & 1 \\ 0 & -1 & 0 & -1 & 0 & -1 & 0 & -1 \end{bmatrix}.$$

This rotation sequence is illustrated in Fig. A.1, where the axis of rotation is $[2 \quad 2 \quad 0]^T$ and the angle of rotation 180°.

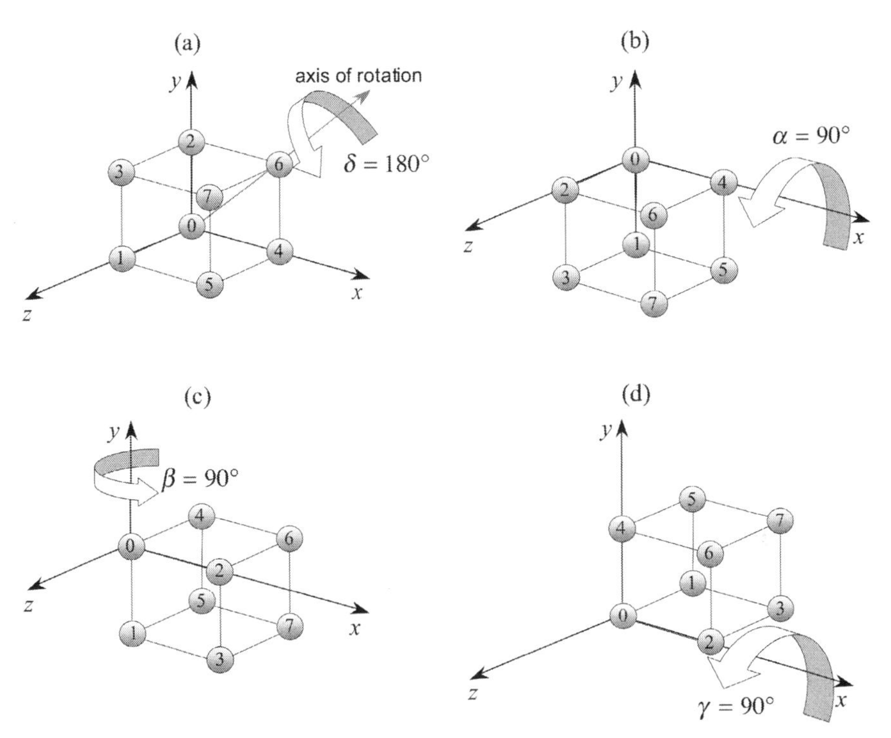


**Fig. A.1** Four views of the unit cube before and during the three rotations $\mathbf{R}_{90°,x}\mathbf{R}_{90°,y}\mathbf{R}_{90°,x}$

## A.3 $\mathbf{R}_{\gamma,x}\mathbf{R}_{\beta,y}\mathbf{R}_{\alpha,z}$

$$\mathbf{R}_{\gamma,x}\mathbf{R}_{\beta,y}\mathbf{R}_{\alpha,z} = \begin{bmatrix} 1 & 0 & 0 \\ 0 & c_\gamma & -s_\gamma \\ 0 & s_\gamma & c_\gamma \end{bmatrix} \begin{bmatrix} c_\beta & 0 & s_\beta \\ 0 & 1 & 0 \\ -s_\beta & 0 & c_\beta \end{bmatrix} \begin{bmatrix} c_\alpha & -s_\alpha & 0 \\ s_\alpha & c_\alpha & 0 \\ 0 & 0 & 1 \end{bmatrix}$$

$$= \begin{bmatrix} c_\beta c_\alpha & -c_\beta s_\alpha & s_\beta \\ (c_\gamma s_\alpha + s_\gamma s_\beta c_\alpha) & (c_\gamma c_\alpha - s_\gamma s_\beta s_\alpha) & -s_\gamma c_\beta \\ (s_\gamma s_\alpha - c_\gamma s_\beta c_\alpha) & (s_\gamma c_\alpha + c_\gamma s_\beta s_\alpha) & c_\gamma c_\beta \end{bmatrix}$$

$$\mathbf{R}_{90°,x}\mathbf{R}_{90°,y}\mathbf{R}_{90°,z} = \begin{bmatrix} 0 & 0 & 1 \\ 0 & -1 & 0 \\ 1 & 0 & 0 \end{bmatrix}$$

$$\begin{bmatrix} 0 & 0 & 1 \\ 0 & -1 & 0 \\ 1 & 0 & 0 \end{bmatrix} \begin{bmatrix} 0 & 0 & 0 & 0 & 1 & 1 & 1 & 1 \\ 0 & 0 & 1 & 1 & 0 & 0 & 1 & 1 \\ 0 & 1 & 0 & 1 & 0 & 1 & 0 & 1 \end{bmatrix}$$

$$= \begin{bmatrix} 0 & 1 & 0 & 1 & 0 & 1 & 0 & 1 \\ 0 & 0 & -1 & -1 & 0 & 0 & -1 & -1 \\ 0 & 0 & 0 & 0 & 1 & 1 & 1 & 1 \end{bmatrix}.$$

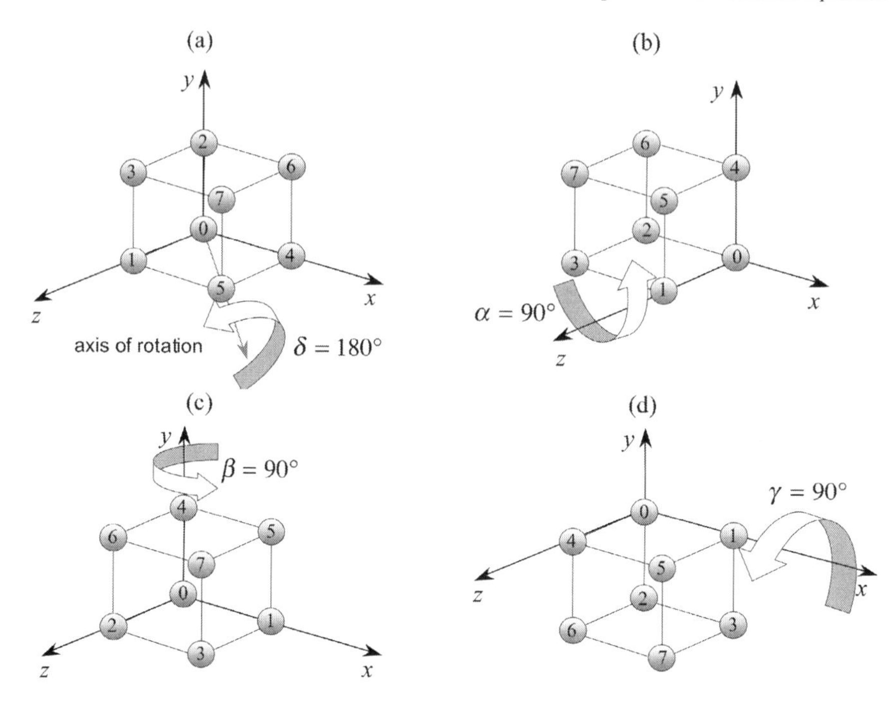


**Fig. A.2** Four views of the unit cube before and during the three rotations $\mathbf{R}_{90^\circ,x}\mathbf{R}_{90^\circ,y}\mathbf{R}_{90^\circ,z}$

This rotation sequence is illustrated in Fig. A.2, where the axis of rotation is $[2 \quad 0 \quad 2]^{\mathrm{T}}$ and the angle of rotation 180°.

## A.4 $\mathbf{R}_{\gamma,x}\mathbf{R}_{\beta,z}\mathbf{R}_{\alpha,x}$

$$\mathbf{R}_{\gamma,x}\mathbf{R}_{\beta,z}\mathbf{R}_{\alpha,x} = \begin{bmatrix} 1 & 0 & 0 \\ 0 & c_\gamma & -s_\gamma \\ 0 & s_\gamma & c_\gamma \end{bmatrix} \begin{bmatrix} c_\beta & -s_\beta & 0 \\ s_\beta & c_\beta & 0 \\ 0 & 0 & 1 \end{bmatrix} \begin{bmatrix} 1 & 0 & 0 \\ 0 & c_\alpha & -s_\alpha \\ 0 & s_\alpha & c_\alpha \end{bmatrix}$$

$$= \begin{bmatrix} c_\beta & -s_\beta c_\alpha & s_\beta s_\alpha \\ c_\gamma s_\beta & (-s_\gamma s_\alpha + c_\gamma c_\beta c_\alpha) & (-s_\gamma c_\alpha - c_\gamma c_\beta s_\alpha) \\ s_\gamma s_\beta & (c_\gamma s_\alpha + s_\gamma c_\beta c_\alpha) & (c_\gamma c_\alpha - s_\gamma c_\beta s_\alpha) \end{bmatrix}$$

$$\mathbf{R}_{90^\circ,x}\mathbf{R}_{90^\circ,z}\mathbf{R}_{90^\circ,x} = \begin{bmatrix} 0 & 0 & 1 \\ 0 & -1 & 0 \\ 1 & 0 & 0 \end{bmatrix}$$

$$\begin{bmatrix} 0 & 0 & 1 \\ 0 & -1 & 0 \\ 1 & 0 & 0 \end{bmatrix} \begin{bmatrix} 0 & 0 & 0 & 0 & 1 & 1 & 1 & 1 \\ 0 & 0 & 1 & 1 & 0 & 0 & 1 & 1 \\ 0 & 1 & 0 & 1 & 0 & 1 & 0 & 1 \end{bmatrix}$$

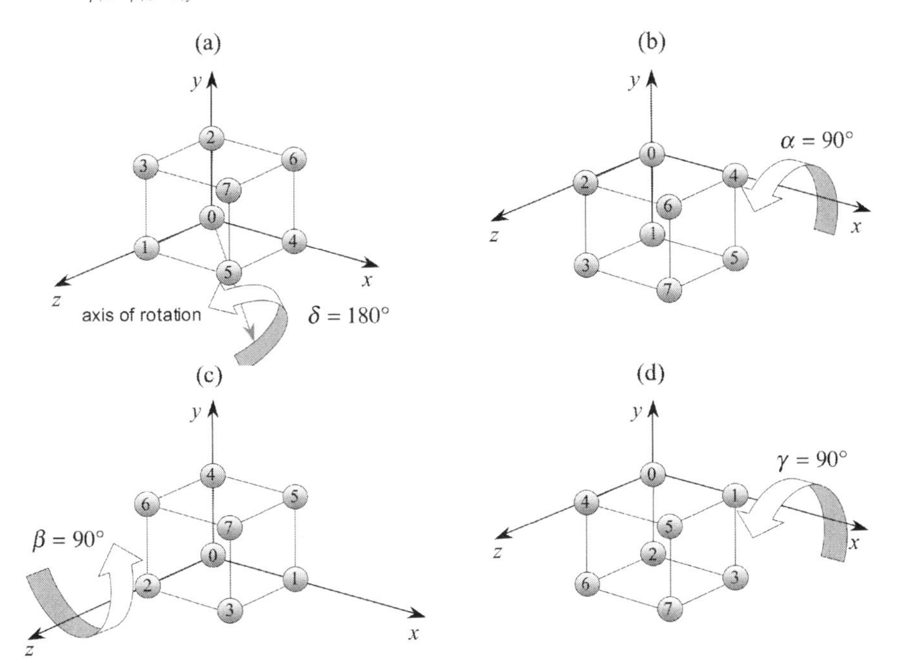


**Fig. A.3** Four views of the unit cube before and during the three rotations $\mathbf{R}_{90°,x}\mathbf{R}_{90°,z}\mathbf{R}_{90°,x}$

$$= \begin{bmatrix} 0 & 1 & 0 & 1 & 0 & 1 & 0 & 1 \\ 0 & 0 & -1 & -1 & 0 & 0 & -1 & -1 \\ 0 & 0 & 0 & 0 & 1 & 1 & 1 & 1 \end{bmatrix}.$$

This rotation sequence is illustrated in Fig. A.3, where the axis of rotation is $[2 \quad 0 \quad 2]^T$ and the angle of rotation 180°.

## A.5 $\mathbf{R}_{\gamma,x}\mathbf{R}_{\beta,z}\mathbf{R}_{\alpha,y}$

$$\mathbf{R}_{\gamma,x}\mathbf{R}_{\beta,z}\mathbf{R}_{\alpha,y} = \begin{bmatrix} 1 & 0 & 0 \\ 0 & c_\gamma & -s_\gamma \\ 0 & s_\gamma & c_\gamma \end{bmatrix} \begin{bmatrix} c_\beta & -s_\beta & 0 \\ s_\beta & c_\beta & 0 \\ 0 & 0 & 1 \end{bmatrix} \begin{bmatrix} c_\alpha & 0 & s_\alpha \\ 0 & 1 & 0 \\ -s_\alpha & 0 & c_\alpha \end{bmatrix}$$

$$= \begin{bmatrix} c_\beta c_\alpha & -s_\beta & c_\beta s_\alpha \\ (s_\gamma s_\alpha + c_\gamma s_\beta c_\alpha) & c_\gamma c_\beta & (-s_\gamma c_\alpha + c_\gamma s_\beta s_\alpha) \\ (-c_\gamma s_\alpha + s_\gamma s_\beta c_\alpha) & s_\gamma c_\beta & (c_\gamma c_\alpha + s_\gamma s_\beta s_\alpha) \end{bmatrix}$$

$$\mathbf{R}_{90°,x}\mathbf{R}_{90°,z}\mathbf{R}_{90°,y} = \begin{bmatrix} 0 & -1 & 0 \\ 1 & 0 & 0 \\ 0 & 0 & 1 \end{bmatrix}$$

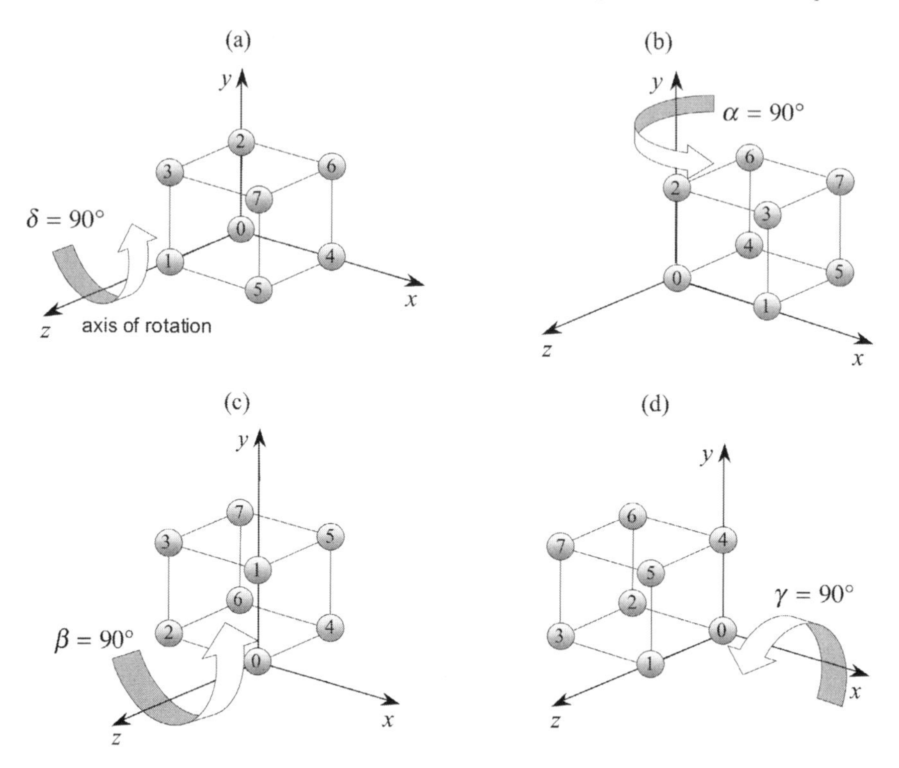


**Fig. A.4** Four views of the unit cube before and during the three rotations $\mathbf{R}_{90°,x}\mathbf{R}_{90°,z}\mathbf{R}_{90°,y}$

$$\begin{bmatrix} 0 & -1 & 0 \\ 1 & 0 & 0 \\ 0 & 0 & 1 \end{bmatrix} \begin{bmatrix} 0 & 0 & 0 & 0 & 1 & 1 & 1 & 1 \\ 0 & 0 & 1 & 1 & 0 & 0 & 1 & 1 \\ 0 & 1 & 0 & 1 & 0 & 1 & 0 & 1 \end{bmatrix}$$
$$= \begin{bmatrix} 0 & 0 & -1 & -1 & 0 & 0 & -1 & -1 \\ 0 & 0 & 0 & 0 & 1 & 1 & 1 & 1 \\ 0 & 1 & 0 & 1 & 0 & 1 & 0 & 1 \end{bmatrix}.$$

This rotation sequence is illustrated in Fig. A.4, where the axis of rotation is $[0 \quad 0 \quad 2]^{\mathrm{T}}$ and the angle of rotation 90°.

## A.6 $\mathbf{R}_{\gamma,y}\mathbf{R}_{\beta,x}\mathbf{R}_{\alpha,y}$

$$\mathbf{R}_{\gamma,y}\mathbf{R}_{\beta,x}\mathbf{R}_{\alpha,y} = \begin{bmatrix} c_\gamma & 0 & s_\gamma \\ 0 & 1 & 0 \\ -s_\gamma & 0 & c_\gamma \end{bmatrix} \begin{bmatrix} 1 & 0 & 0 \\ 0 & c_\beta & -s_\beta \\ 0 & s_\beta & c_\beta \end{bmatrix} \begin{bmatrix} c_\alpha & 0 & s_\alpha \\ 0 & 1 & 0 \\ -s_\alpha & 0 & c_\alpha \end{bmatrix}$$

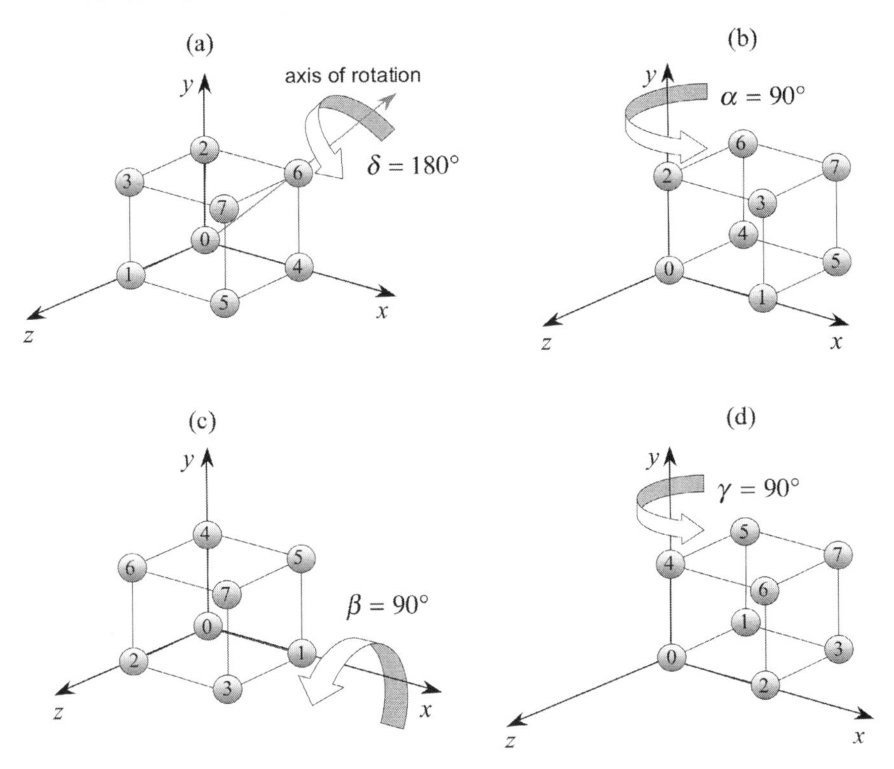


**Fig. A.5** Four views of the unit cube before and during the three rotations $\mathbf{R}_{90^\circ,y}\mathbf{R}_{90^\circ,x}\mathbf{R}_{90^\circ,y}$

$$= \begin{bmatrix} (c_\gamma c_\alpha - s_\gamma c_\beta s_\alpha) & s_\gamma s_\beta & (c_\gamma s_\alpha + s_\gamma c_\beta c_\alpha) \\ s_\beta s_\alpha & c_\beta & -s_\beta c_\alpha \\ (-s_\gamma c_\alpha - c_\gamma c_\beta s_\alpha) & c_\gamma s_\beta & (-s_\gamma s_\alpha + c_\gamma c_\beta c_\alpha) \end{bmatrix}$$

$$\mathbf{R}_{90^\circ,y}\mathbf{R}_{90^\circ,x}\mathbf{R}_{90^\circ,y} = \begin{bmatrix} 0 & 1 & 0 \\ 1 & 0 & 0 \\ 0 & 0 & -1 \end{bmatrix}$$

$$\begin{bmatrix} 0 & 1 & 0 \\ 1 & 0 & 0 \\ 0 & 0 & -1 \end{bmatrix} \begin{bmatrix} 0 & 0 & 0 & 0 & 1 & 1 & 1 & 1 \\ 0 & 0 & 1 & 1 & 0 & 0 & 1 & 1 \\ 0 & 1 & 0 & 1 & 0 & 1 & 0 & 1 \end{bmatrix}$$
$$= \begin{bmatrix} 0 & 0 & 1 & 1 & 0 & 0 & 1 & 1 \\ 0 & 0 & 0 & 0 & 1 & 1 & 1 & 1 \\ 0 & -1 & 0 & -1 & 0 & -1 & 0 & -1 \end{bmatrix}.$$

This rotation sequence is illustrated in Fig. A.5, where the axis of rotation is $[2 \quad 2 \quad 0]^{\mathrm{T}}$ and the angle of rotation 180°.

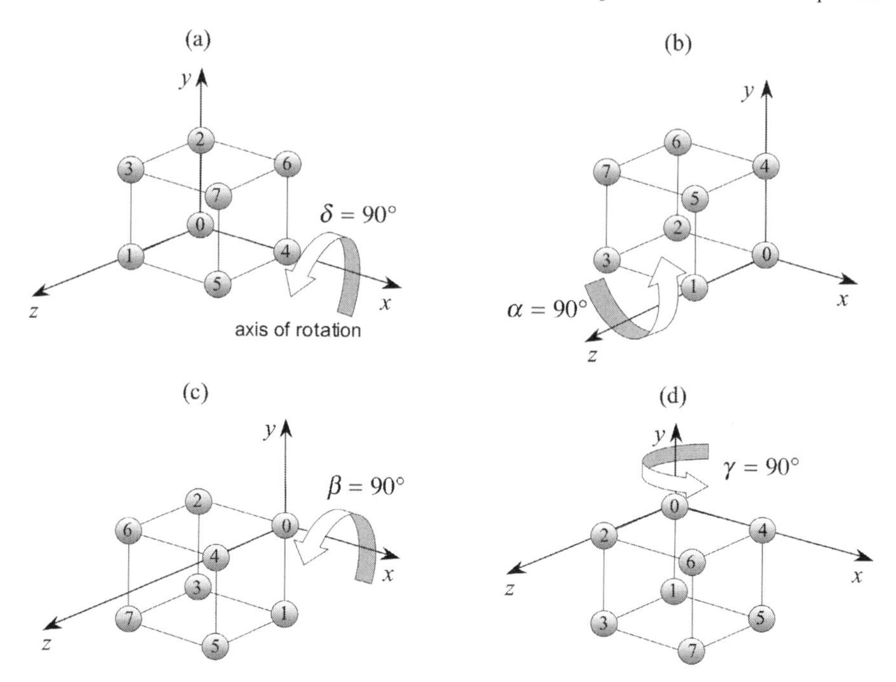


**Fig. A.6** Four views of the unit cube before and during the three rotations $\mathbf{R}_{90^\circ,y}\mathbf{R}_{90^\circ,x}\mathbf{R}_{90^\circ,z}$

## A.7 $\mathbf{R}_{\gamma,y}\mathbf{R}_{\beta,x}\mathbf{R}_{\alpha,z}$

$$\mathbf{R}_{\gamma,y}\mathbf{R}_{\beta,x}\mathbf{R}_{\alpha,z} = \begin{bmatrix} c_\gamma & 0 & s_\gamma \\ 0 & 1 & 0 \\ -s_\gamma & 0 & c_\gamma \end{bmatrix}\begin{bmatrix} 1 & 0 & 0 \\ 0 & c_\beta & -s_\beta \\ 0 & s_\beta & c_\beta \end{bmatrix}\begin{bmatrix} c_\alpha & -s_\alpha & 0 \\ s_\alpha & c_\alpha & 0 \\ 0 & 0 & 1 \end{bmatrix}$$

$$= \begin{bmatrix} (c_\gamma c_\alpha + s_\gamma s_\beta s_\alpha) & (-c_\gamma s_\alpha + s_\gamma s_\beta c_\alpha) & s_\gamma c_\beta \\ c_\beta s_\alpha & c_\beta c_\alpha & -s_\beta \\ (-s_\gamma c_\alpha + c_\gamma s_\beta s_\alpha) & (s_\gamma s_\alpha + c_\gamma s_\beta c_\alpha) & c_\gamma c_\beta \end{bmatrix}$$

$$\mathbf{R}_{90^\circ,y}\mathbf{R}_{90^\circ,x}\mathbf{R}_{90^\circ,z} = \begin{bmatrix} 1 & 0 & 0 \\ 0 & 0 & -1 \\ 0 & 1 & 0 \end{bmatrix}$$

$$\begin{bmatrix} 1 & 0 & 0 \\ 0 & 0 & -1 \\ 0 & 1 & 0 \end{bmatrix}\begin{bmatrix} 0 & 0 & 0 & 0 & 1 & 1 & 1 & 1 \\ 0 & 0 & 1 & 1 & 0 & 0 & 1 & 1 \\ 0 & 1 & 0 & 1 & 0 & 1 & 0 & 1 \end{bmatrix}$$

$$= \begin{bmatrix} 0 & 0 & 0 & 0 & 1 & 1 & 1 & 1 \\ 0 & -1 & 0 & -1 & 0 & -1 & 0 & -1 \\ 0 & 0 & 1 & 1 & 0 & 0 & 1 & 1 \end{bmatrix}.$$

This rotation sequence is illustrated in Fig. A.6, where the axis of rotation is $[2 \quad 0 \quad 0]^{\mathrm{T}}$ and the angle of rotation 90°.

## A.8 $\mathbf{R}_{\gamma,y}\mathbf{R}_{\beta,z}\mathbf{R}_{\alpha,x}$

$$\mathbf{R}_{\gamma,y}\mathbf{R}_{\beta,z}\mathbf{R}_{\alpha,x} = \begin{bmatrix} c_\gamma & 0 & s_\gamma \\ 0 & 1 & 0 \\ -s_\gamma & 0 & c_\gamma \end{bmatrix} \begin{bmatrix} c_\beta & -s_\beta & 0 \\ s_\beta & c_\beta & 0 \\ 0 & 0 & 1 \end{bmatrix} \begin{bmatrix} 1 & 0 & 0 \\ 0 & c_\alpha & -s_\alpha \\ 0 & s_\alpha & c_\alpha \end{bmatrix}$$

$$= \begin{bmatrix} c_\gamma c_\beta & (s_\gamma s_\alpha - c_\gamma s_\beta c_\alpha) & (s_\gamma c_\alpha + c_\gamma s_\beta s_\alpha) \\ s_\beta & c_\beta c_\alpha & -c_\beta s_\alpha \\ -s_\gamma c_\beta & (c_\gamma s_\alpha + s_\gamma s_\beta c_\alpha) & (c_\gamma c_\alpha - s_\gamma s_\beta s_\alpha) \end{bmatrix}$$

$$\mathbf{R}_{90^\circ,y}\mathbf{R}_{90^\circ,z}\mathbf{R}_{90^\circ,x} = \begin{bmatrix} 0 & 1 & 0 \\ 1 & 0 & 0 \\ 0 & 0 & -1 \end{bmatrix}$$

$$\begin{bmatrix} 0 & 1 & 0 \\ 1 & 0 & 0 \\ 0 & 0 & -1 \end{bmatrix} \begin{bmatrix} 0 & 0 & 0 & 0 & 1 & 1 & 1 & 1 \\ 0 & 0 & 1 & 1 & 0 & 0 & 1 & 1 \\ 0 & 1 & 0 & 1 & 0 & 1 & 0 & 1 \end{bmatrix}$$
$$= \begin{bmatrix} 0 & 0 & 1 & 1 & 0 & 0 & 1 & 1 \\ 0 & 0 & 0 & 0 & 1 & 1 & 1 & 1 \\ 0 & -1 & 0 & -1 & 0 & -1 & 0 & -1 \end{bmatrix}.$$

This rotation sequence is illustrated in Fig. A.7, where the axis of rotation is $[2 \quad 2 \quad 0]^T$ and the angle of rotation 180°.

## A.9 $\mathbf{R}_{\gamma,y}\mathbf{R}_{\beta,z}\mathbf{R}_{\alpha,y}$

$$\mathbf{R}_{\gamma,y}\mathbf{R}_{\beta,z}\mathbf{R}_{\alpha,y} = \begin{bmatrix} c_\gamma & 0 & s_\gamma \\ 0 & 1 & 0 \\ -s_\gamma & 0 & c_\gamma \end{bmatrix} \begin{bmatrix} c_\beta & -s_\beta & 0 \\ s_\beta & c_\beta & 0 \\ 0 & 0 & 1 \end{bmatrix} \begin{bmatrix} c_\alpha & 0 & s_\alpha \\ 0 & 1 & 0 \\ -s_\alpha & 0 & c_\alpha \end{bmatrix}$$

$$= \begin{bmatrix} (-s_\gamma s_\alpha + c_\gamma c_\beta c_\alpha) & -c_\gamma s_\beta & (s_\gamma c_\alpha + c_\gamma c_\beta s_\alpha) \\ s_\beta c_\alpha & c_\beta & s_\beta s_\alpha \\ (-c_\gamma s_\alpha - s_\gamma c_\beta c_\alpha) & s_\gamma s_\beta & (c_\gamma c_\alpha - s_\gamma c_\beta s_\alpha) \end{bmatrix}$$

$$\mathbf{R}_{90^\circ,y}\mathbf{R}_{90^\circ,z}\mathbf{R}_{90^\circ,y} = \begin{bmatrix} -1 & 0 & 0 \\ 0 & 0 & 1 \\ 0 & 1 & 0 \end{bmatrix}$$

$$\begin{bmatrix} -1 & 0 & 0 \\ 0 & 0 & 1 \\ 0 & 1 & 0 \end{bmatrix} \begin{bmatrix} 0 & 0 & 0 & 0 & 1 & 1 & 1 & 1 \\ 0 & 0 & 1 & 1 & 0 & 0 & 1 & 1 \\ 0 & 1 & 0 & 1 & 0 & 1 & 0 & 1 \end{bmatrix}$$

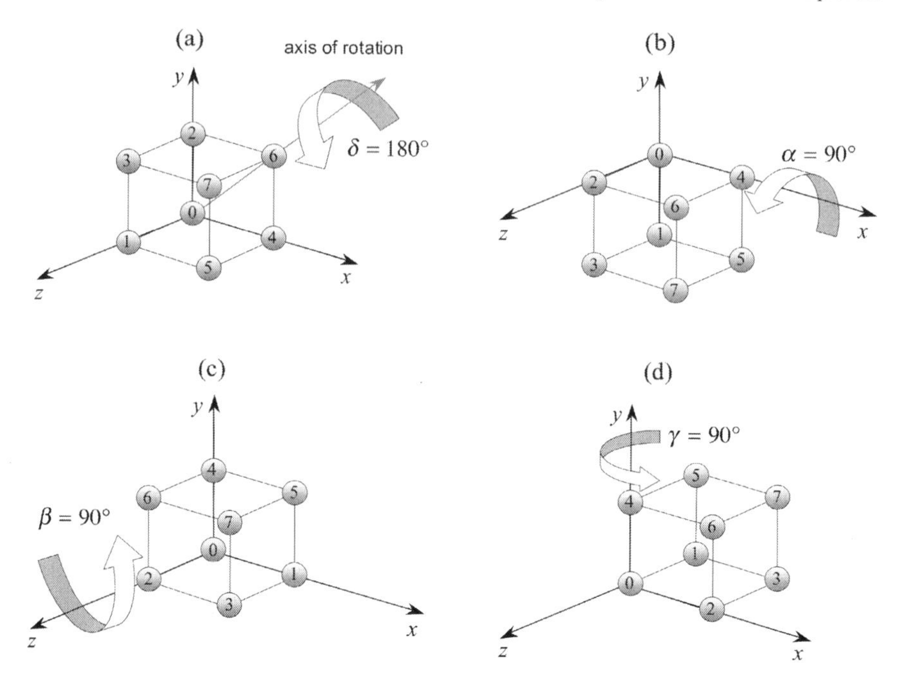


**Fig. A.7** Four views of the unit cube before and during the three rotations $\mathbf{R}_{90°,y}\mathbf{R}_{90°,z}\mathbf{R}_{90°,x}$

$$= \begin{bmatrix} 0 & 0 & 0 & 0 & -1 & -1 & -1 & -1 \\ 0 & 1 & 0 & 1 & 0 & 1 & 0 & 1 \\ 0 & 0 & 1 & 1 & 0 & 0 & 1 & 1 \end{bmatrix}.$$

This rotation sequence is illustrated in Fig. A.8, where the axis of rotation is $[0 \quad 2 \quad 2]^{\mathrm{T}}$ and the angle of rotation 180°.

## A.10 $\mathbf{R}_{\gamma,z}\mathbf{R}_{\beta,x}\mathbf{R}_{\alpha,y}$

$$\mathbf{R}_{\gamma,z}\mathbf{R}_{\beta,x}\mathbf{R}_{\alpha,y} = \begin{bmatrix} c_\gamma & -s_\gamma & 0 \\ s_\gamma & c_\gamma & 0 \\ 0 & 0 & 1 \end{bmatrix} \begin{bmatrix} 1 & 0 & 0 \\ 0 & c_\beta & -s_\beta \\ 0 & s_\beta & c_\beta \end{bmatrix} \begin{bmatrix} c_\alpha & 0 & s_\alpha \\ 0 & 1 & 0 \\ -s_\alpha & 0 & c_\alpha \end{bmatrix}$$

$$= \begin{bmatrix} (c_\gamma c_\alpha - s_\gamma s_\beta s_\alpha) & -s_\gamma c_\beta & (c_\gamma s_\alpha + s_\gamma s_\beta c_\alpha) \\ (s_\gamma c_\alpha + c_\gamma s_\beta s_\alpha) & c_\gamma c_\beta & (s_\gamma s_\alpha - c_\gamma s_\beta c_\alpha) \\ -c_\beta s_\alpha & s_\beta & c_\beta c_\alpha \end{bmatrix}$$

$$\mathbf{R}_{90°,z}\mathbf{R}_{90°,x}\mathbf{R}_{90°,y} = \begin{bmatrix} -1 & 0 & 0 \\ 0 & 0 & 1 \\ 0 & 1 & 0 \end{bmatrix}$$

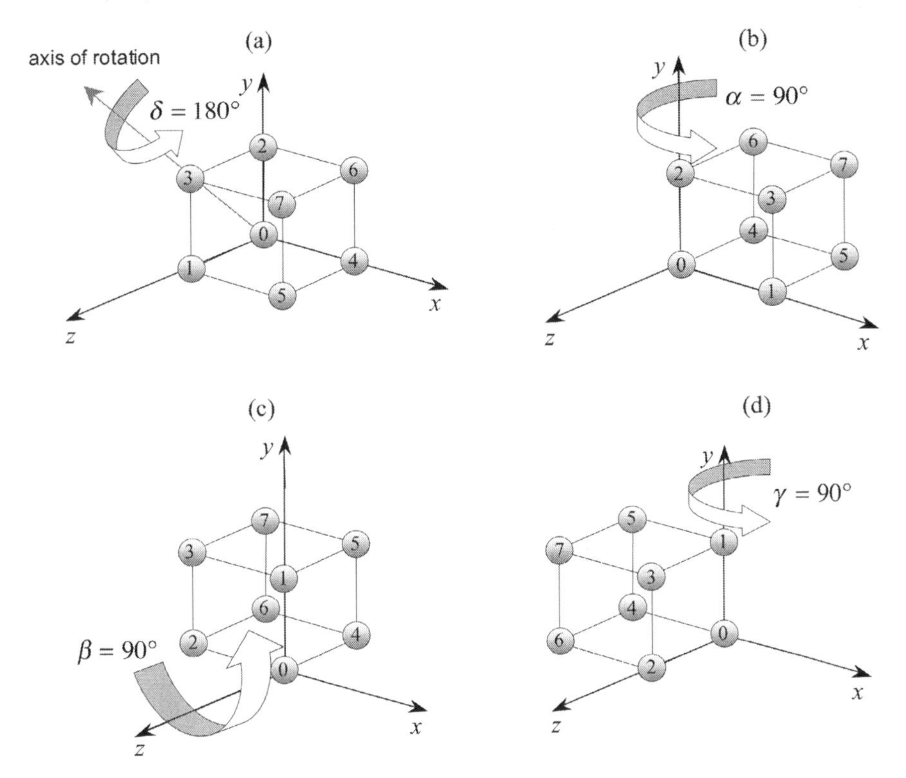


**Fig. A.8** Four views of the unit cube before and during the three rotations $\mathbf{R}_{90^\circ,y}\mathbf{R}_{90^\circ,z}\mathbf{R}_{90^\circ,y}$

$$\begin{bmatrix} -1 & 0 & 0 \\ 0 & 0 & 1 \\ 0 & 1 & 0 \end{bmatrix}\begin{bmatrix} 0 & 0 & 0 & 0 & 1 & 1 & 1 & 1 \\ 0 & 0 & 1 & 1 & 0 & 0 & 1 & 1 \\ 0 & 1 & 0 & 1 & 0 & 1 & 0 & 1 \end{bmatrix}$$
$$= \begin{bmatrix} 0 & 0 & 0 & 0 & -1 & -1 & -1 & -1 \\ 0 & 1 & 0 & 1 & 0 & 1 & 0 & 1 \\ 0 & 0 & 1 & 1 & 0 & 0 & 1 & 1 \end{bmatrix}.$$

This rotation sequence is illustrated in Fig. A.9, where the axis of rotation is $[0 \quad 2 \quad 2]^T$ and the angle of rotation 180°.

## A.11 $\mathbf{R}_{\gamma,z}\mathbf{R}_{\beta,x}\mathbf{R}_{\alpha,z}$

$$\mathbf{R}_{\gamma,z}\mathbf{R}_{\beta,x}\mathbf{R}_{\alpha,z} = \begin{bmatrix} c_\gamma & -s_\gamma & 0 \\ s_\gamma & c_\gamma & 0 \\ 0 & 0 & 1 \end{bmatrix}\begin{bmatrix} 1 & 0 & 0 \\ 0 & c_\beta & -s_\beta \\ 0 & s_\beta & c_\beta \end{bmatrix}\begin{bmatrix} c_\alpha & -s_\alpha & 0 \\ s_\alpha & c_\alpha & 0 \\ 0 & 0 & 1 \end{bmatrix}$$

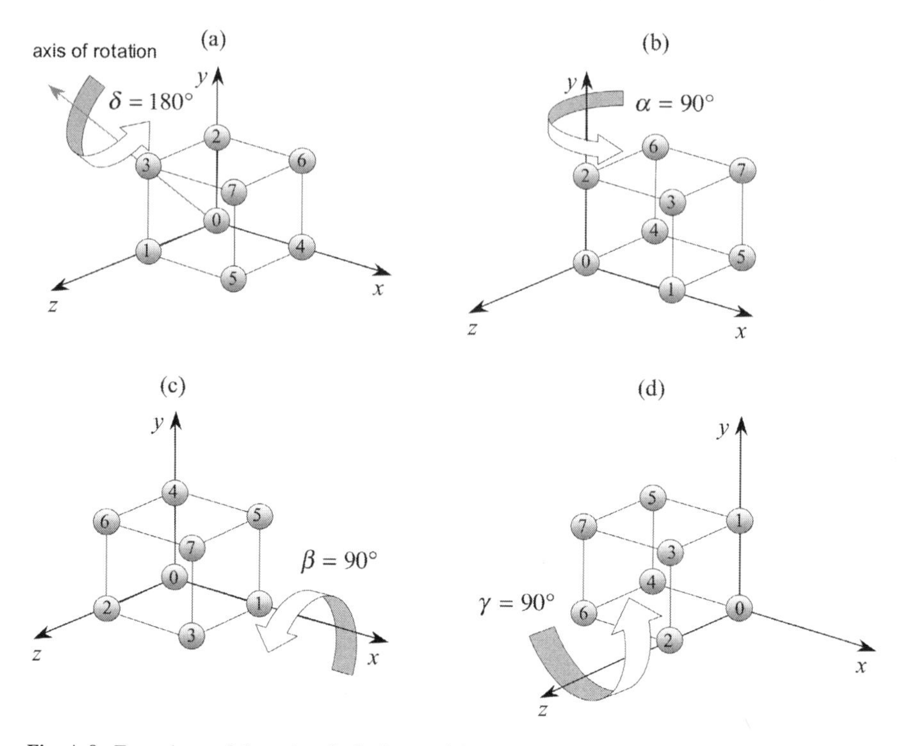


**Fig. A.9** Four views of the unit cube before and during the three rotations $\mathbf{R}_{90°,z}\mathbf{R}_{90°,x}\mathbf{R}_{90°,y}$

$$= \begin{bmatrix} (c_\gamma c_\alpha - s_\gamma c_\beta s_\alpha) & (-c_\gamma s_\alpha - s_\gamma c_\beta c_\alpha) & s_\gamma s_\beta \\ (s_\gamma c_\alpha + c_\gamma c_\beta s_\alpha) & (-s_\gamma s_\alpha + c_\gamma c_\beta c_\alpha) & -c_\gamma s_\beta \\ s_\beta s_\alpha & s_\beta c_\alpha & c_\beta \end{bmatrix}$$

$$\mathbf{R}_{90°,z}\mathbf{R}_{90°,x}\mathbf{R}_{90°,z} = \begin{bmatrix} 0 & 0 & 1 \\ 0 & -1 & 0 \\ 1 & 0 & 0 \end{bmatrix}$$

$$\begin{bmatrix} 0 & 0 & 1 \\ 0 & -1 & 0 \\ 1 & 0 & 0 \end{bmatrix} \begin{bmatrix} 0 & 0 & 0 & 0 & 1 & 1 & 1 & 1 \\ 0 & 0 & 1 & 1 & 0 & 0 & 1 & 1 \\ 0 & 1 & 0 & 1 & 0 & 1 & 0 & 1 \end{bmatrix}$$
$$= \begin{bmatrix} 0 & 1 & 0 & 1 & 0 & 1 & 0 & 1 \\ 0 & 0 & -1 & -1 & 0 & 0 & -1 & -1 \\ 0 & 0 & 0 & 0 & 1 & 1 & 1 & 1 \end{bmatrix}.$$

This rotation sequence is illustrated in Fig. A.10, where the axis of rotation is $[2 \quad 0 \quad 2]^{\mathrm{T}}$ and the angle of rotation 180°.

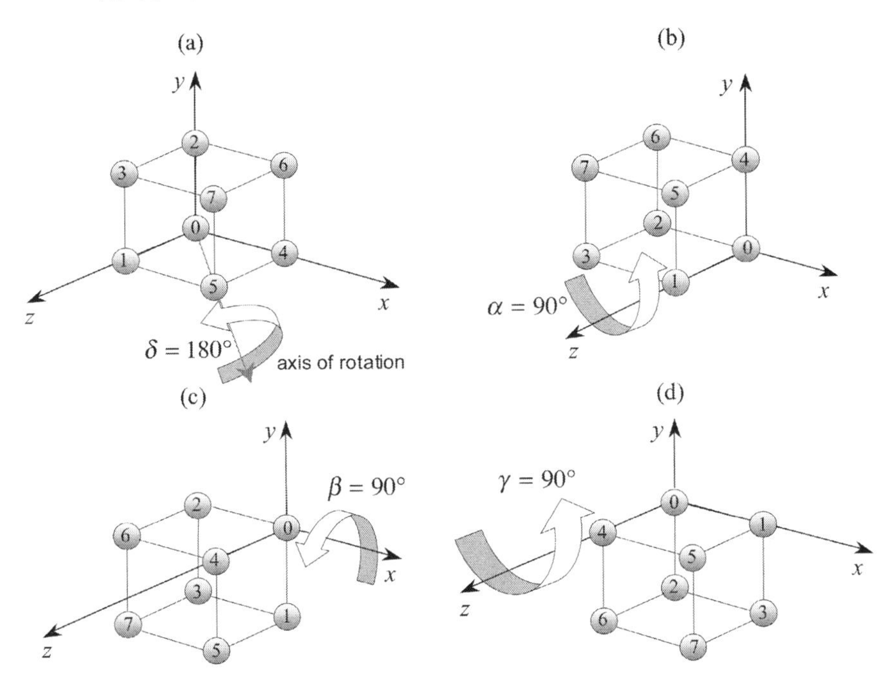


**Fig. A.10** Four views of the unit cube before and during the three rotations $\mathbf{R}_{90°,z}\mathbf{R}_{90°,x}\mathbf{R}_{90°,z}$

## A.12 $\mathbf{R}_{\gamma,z}\mathbf{R}_{\beta,y}\mathbf{R}_{\alpha,x}$

$$\mathbf{R}_{\gamma,z}\mathbf{R}_{\beta,y}\mathbf{R}_{\alpha,x} = \begin{bmatrix} c_\gamma & -s_\gamma & 0 \\ s_\gamma & c_\gamma & 0 \\ 0 & 0 & 1 \end{bmatrix} \begin{bmatrix} c_\beta & 0 & s_\beta \\ 0 & 1 & 0 \\ -s_\beta & 0 & c_\beta \end{bmatrix} \begin{bmatrix} 1 & 0 & 0 \\ 0 & c_\alpha & -s_\alpha \\ 0 & s_\alpha & c_\alpha \end{bmatrix}$$

$$= \begin{bmatrix} c_\gamma c_\beta & (-s_\gamma c_\alpha + c_\gamma s_\beta s_\alpha) & (s_\gamma s_\alpha + c_\gamma s_\beta c_\alpha) \\ s_\gamma c_\beta & (c_\gamma c_\alpha + s_\gamma s_\beta s_\alpha) & (-c_\gamma s_\alpha + s_\gamma s_\beta c_\alpha) \\ -s_\beta & c_\beta s_\alpha & c_\beta c_\alpha) \end{bmatrix}$$

$$\mathbf{R}_{90°,z}\mathbf{R}_{90°,y}\mathbf{R}_{90°,x} = \begin{bmatrix} 0 & 0 & 1 \\ 0 & 1 & 0 \\ -1 & 0 & 0 \end{bmatrix}$$

$$\begin{bmatrix} 0 & 0 & 1 \\ 0 & 1 & 0 \\ -1 & 0 & 0 \end{bmatrix} \begin{bmatrix} 0 & 0 & 0 & 0 & 1 & 1 & 1 & 1 \\ 0 & 0 & 1 & 1 & 0 & 0 & 1 & 1 \\ 0 & 1 & 0 & 1 & 0 & 1 & 0 & 1 \end{bmatrix}$$

$$= \begin{bmatrix} 0 & 1 & 0 & 1 & 0 & 1 & 0 & 1 \\ 0 & 0 & 1 & 1 & 0 & 0 & 1 & 1 \\ 0 & 0 & 0 & 0 & -1 & -1 & -1 & -1 \end{bmatrix}.$$

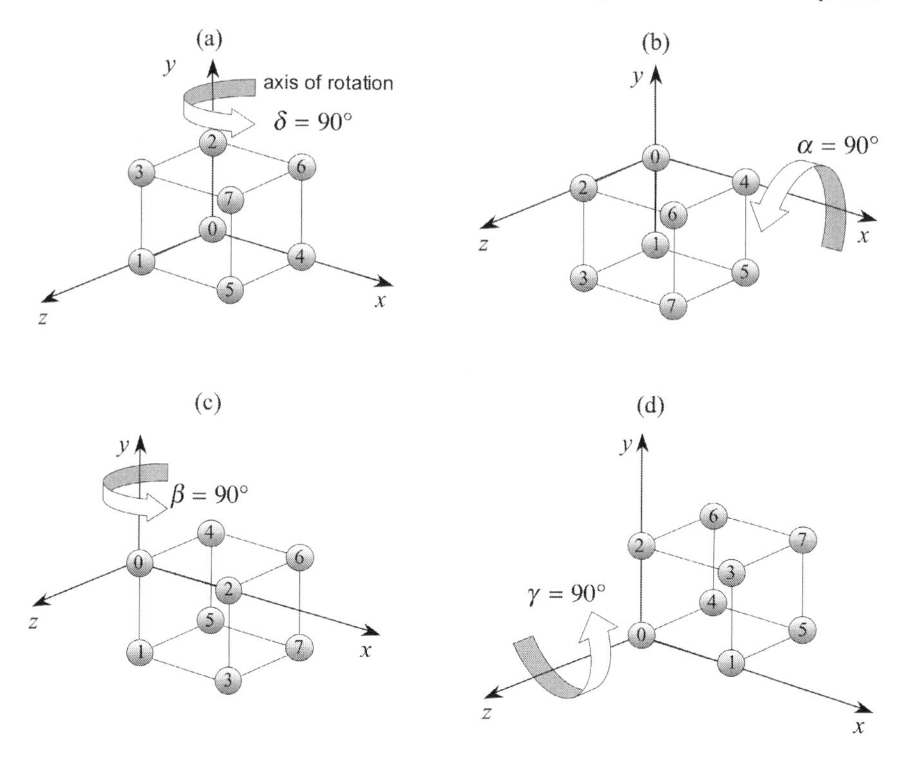


**Fig. A.11** Four views of the unit cube before and during the three rotations $\mathbf{R}_{90^\circ,z}\mathbf{R}_{90^\circ,y}\mathbf{R}_{90^\circ,x}$

This rotation sequence is illustrated in Fig. A.11, where the axis of rotation is $[0 \quad 2 \quad 0]^{\mathrm{T}}$ and the angle of rotation 90°.

## A.13 $\mathbf{R}_{\gamma,z}\mathbf{R}_{\beta,y}\mathbf{R}_{\alpha,z}$

$$\mathbf{R}_{\gamma,z}\mathbf{R}_{\beta,y}\mathbf{R}_{\alpha,z} = \begin{bmatrix} c_\gamma & -s_\gamma & 0 \\ s_\gamma & c_\gamma & 0 \\ 0 & 0 & 1 \end{bmatrix}\begin{bmatrix} c_\beta & 0 & s_\beta \\ 0 & 1 & 0 \\ -s_\beta & 0 & c_\beta \end{bmatrix}\begin{bmatrix} c_\alpha & -s_\alpha & 0 \\ s_\alpha & c_\alpha & 0 \\ 0 & 0 & 1 \end{bmatrix}$$

$$= \begin{bmatrix} (-s_\gamma s_\alpha + c_\gamma c_\beta c_\alpha) & (-s_\gamma c_\alpha - c_\gamma c_\beta s_\alpha) & c_\gamma s_\beta \\ (c_\gamma s_\alpha + s_\gamma c_\beta c_\alpha) & (c_\gamma c_\alpha - s_\gamma c_\beta s_\alpha) & s_\gamma s_\beta \\ -s_\beta c_\alpha & s_\beta s_\alpha & c_\beta \end{bmatrix}$$

$$\mathbf{R}_{90^\circ,z}\mathbf{R}_{90^\circ,y}\mathbf{R}_{90^\circ,z} = \begin{bmatrix} -1 & 0 & 0 \\ 0 & 0 & 1 \\ 0 & 1 & 0 \end{bmatrix}$$

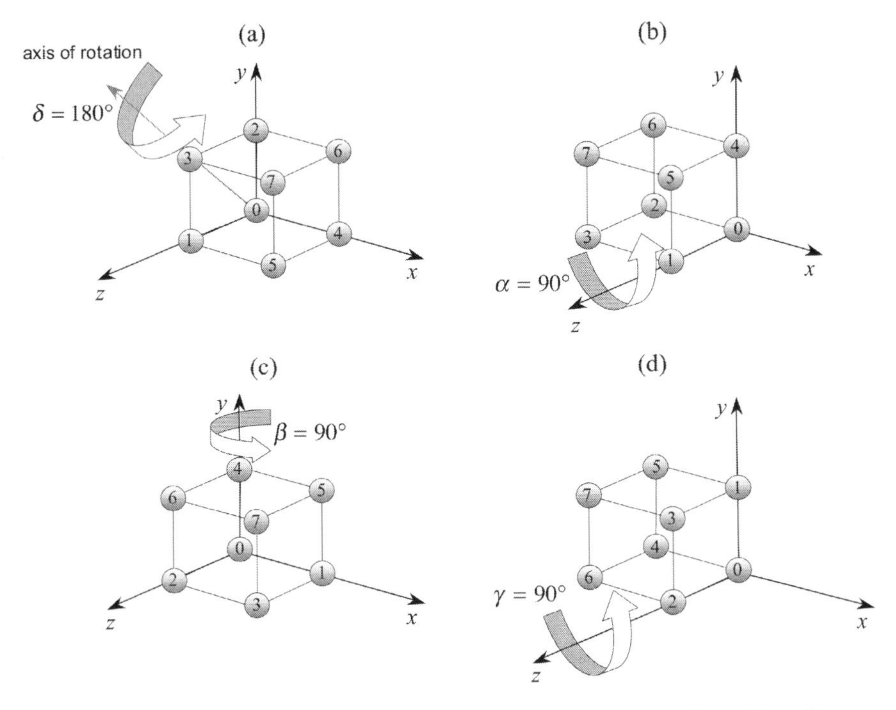


**Fig. A.12** Four views of the unit cube before and during the three rotations $\mathbf{R}_{90°,z}\mathbf{R}_{90°,y}\mathbf{R}_{90°,z}$

$$\begin{bmatrix} -1 & 0 & 0 \\ 0 & 0 & 1 \\ 0 & 1 & 0 \end{bmatrix} \begin{bmatrix} 0 & 0 & 0 & 0 & 1 & 1 & 1 & 1 \\ 0 & 0 & 1 & 1 & 0 & 0 & 1 & 1 \\ 0 & 1 & 0 & 1 & 0 & 1 & 0 & 1 \end{bmatrix}$$
$$= \begin{bmatrix} 0 & 0 & 0 & 0 & -1 & -1 & -1 & -1 \\ 0 & 1 & 0 & 1 & 0 & 1 & 0 & 1 \\ 0 & 0 & 1 & 1 & 0 & 0 & 1 & 1 \end{bmatrix}.$$

This rotation sequence is illustrated in Fig. A.12, where the axis of rotation is $[0 \quad 2 \quad 2]^T$ and the angle of rotation 180°.

# Index



J. Vince, *Matrix Transforms for Computer Games and Animation*,
DOI 10.1007/978-1-4471-4321-5, © Springer-Verlag London 2012